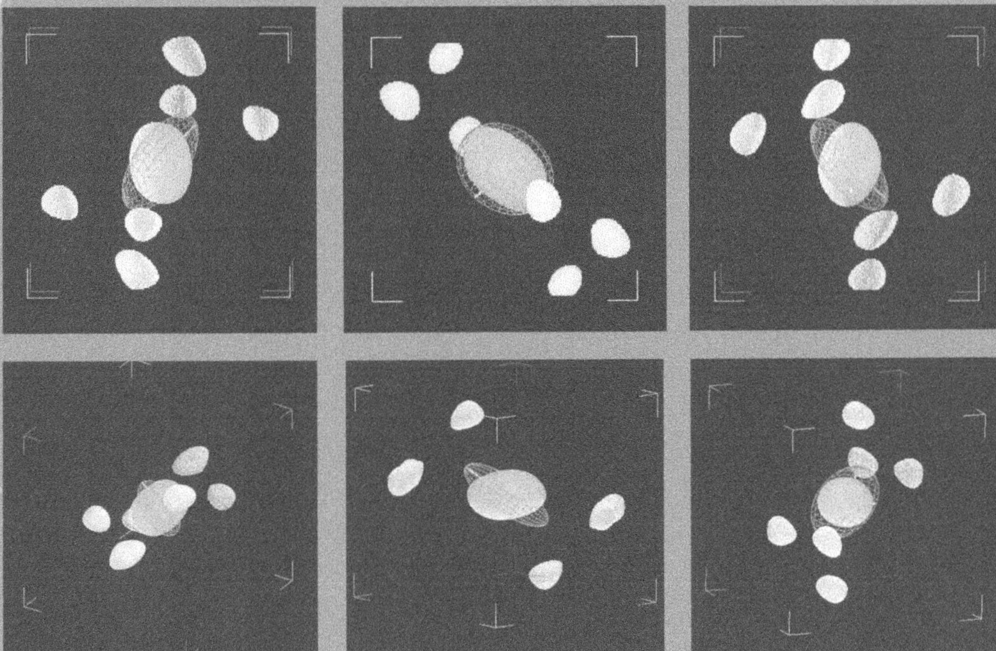

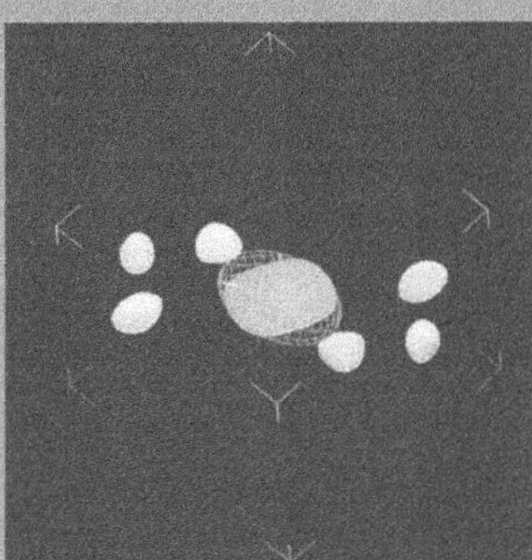

The Difference Electron Nanoscope

The Difference Electron Nanoscope

Methods and Applications

Werner Lottermoser

Pan Stanford Publishing

Published by

Pan Stanford Publishing Pte. Ltd.
Penthouse Level, Suntec Tower 3
8 Temasek Boulevard
Singapore 038988

Email: editorial@panstanford.com
Web: www.panstanford.com

British Library Cataloguing-in-Publication Data
A catalogue record for this book is available from the British Library.

The Difference Electron Nanoscope: Methods and Applications

ISBN 978-981-4774-01-7 (Hardcover)
ISBN 978-1-315-19664-0 (eBook)

MIX
Paper from responsible sources
FSC® C011748

Printed by Ashford Colour Press Ltd

Contents

Preface

The development of the software for the difference electron nanoscope (DEN) has been a long-term collaboration among different authors. The first to be named is my former thesis student Konrad Steiner, who has, in the frame of a project granted by the Austrian Fund of Scientific Research (FWF), provided the preparing program EVOX to transform crystallographic data into voxels to be displayed by the DEN. The co-worker at this stage was the mathematician Hannes Leeb, who has implemented a special FFT (fast Fourier transform) to get a tremendous spare of CPU time. For the programming of the DEN software itself, Gerhard Scharfetter should be named—in collaboration with KS and me, he has developed under the famous special language IDL (interactive data language) from RSI™ a wonderful GUI (graphical user interface) to have a very comfortable access for data input and output on the screen. Moreover, he has programmed the internal calculations, e.g. the integration of the difference electron density hyperareas or the efg (electric field gradient). All these colleagues, however, left my working group and found new jobs, might it be in the industry, the teaching area or other universities as it often happens. However, I am very much indebted to these friends and colleagues—my main duty was to keep the programs running and find new applications thus prohibiting the common work to fall into oblivion. Further, I thank Apple Inc. and the late Steve Jobs to give us a machine at hand that was awesome to the time when the software was created (G5™ with double processor under Mac OS X 10.4 (Tiger)™). It helped us to remain untroubled by system crashes and other boring things that another operating system is "famous" for (I won't explicate this topic). I also owe my deep thanks to my publisher, Stanford Chong, who has taken the risk to promote a book from my very special topic, as the latter may seem at first glance. I hope that I can prove in my present work that applications can touch many fields of the scientific area.

Last but not least, I would like to thank my family for supporting my lone trips—writing of a book is somehow like a first ascent. In this sense I am indebted to my kind daughter, who accompanied me to this memorable Congress of Materials Science 2014 at San Antonio, where I have had the honour to give a talk about the topic of the book and where the initial spark was ignited—she has rendered me luck.

Introduction: What Is a DEN

For the first time in physics history diffractometry and spectroscopy have been joined to common synergetic effects that may contribute to a better understanding of electric and magnetic interactions in a crystal. The DEN (difference electron nanoscope) is no classical machine as the title of the book might infer. It is a smart combination of a very powerful computer and a software that combines diffractometric and spectroscopic data in order to visualize three-dimensional difference electron densities around the nuclei of the ions in the crystal unit cell—in our case stemming from 3d orbitals. These serve to calculate the common quantity of diffractometry and spectroscopy, the electric field gradient (efg) with high accuracy and to display it as a wire frame model within the unit cell of the sample involved. In this sense the DEN is acting on a sub-nanometre scale (therefore the term "nanoscope") and generating images of uncompared symmetrical and physical evidence—and beauty. This will be shown in Chapter 5 of the present book. The experimental efg, however, can be determined from single crystal Mössbauer spectroscopy (SCMBS) or nuclear magnetic/quadrupole resonance (NMR/NQR) that is confined to other nuclides as the SCMBS. Hence, the DEN can be applied to a huge multitude of scientifically interesting specimen since the main method involved, diffractometry in a wide sense, has no general limitations at all. So, the DEN is a rather universal method and the present monograph might contribute to a wide distribution of the method in the scientific world. Has anyone seen a real orbital before? I don't think of the very esthetic, but calculated multipole images to be admired in the common textbooks. I mean: a real orbital distribution in a crystal unit cell together with its efg tensor ellipsoid? In the present monograph, one can see it.

Chapter 1

An Overview on the Methods Involved

The input of the DEN is based on structure factors stemming from X-ray, synchrotron or neutron diffractometry (cf. Sections 3.3.2, 3.3.3, 3.3.4). The aim is to gain an electric field gradient (cf. Chapter 2) that is consistent with the experimental one from single crystal Mössbauer spectroscopy SCMBS (cf. Sections 3.1.1 and 3.1.2) or nuclear magnetic/quadrupole resonance NMR/NQR (cf. Section 3.1.3) and the theoretical one from density functional theory DFT (cf. Section 3.2). Between all these methods different interactions exist, so it seems to be adequate to provide a graphical representation here (Fig. 1.1).

The central quantity and common link of all the applied methods, as emphasized in Fig. 1.1, is the electric field gradient efg that will be given full reference in Chapter 2. As mentioned above, it is threefold: an experimental, a full-quantitative and a semi-quantitative one. The efgs are derived for the special compound under consideration; hence, it is adequate to draw an arrow from the bottom of Fig. 1.1 to the top since the two basic scientific fields (spectroscopy and diffractometry) clearly need a sample (S in Fig. 1.1) to be applied on. All the three efgs mentioned above, however, should be identical concerning size and orientation within error bars. If this is warranted, the relevant data can be published (plots of the difference electron densities within the sample unit cell from different viewpoints

The Difference Electron Nanoscope: Methods and Applications
Werner Lottermoser

ISBN 978-981-4774-01-7 (Hardcover), 978-1-315-19664-0 (eBook)
www.panstanford.com

and directions, the efg as tensor ellipsoid, direction of the magnetic moments from SCMBS and neutron diffraction a.s.o.). These plots are not only interesting from an esthetic point of view: The quasi-three-dimensional character of the images allows to gain an invaluable insight into the main electronic interactions within the sample unit cell that are commonly not visible.

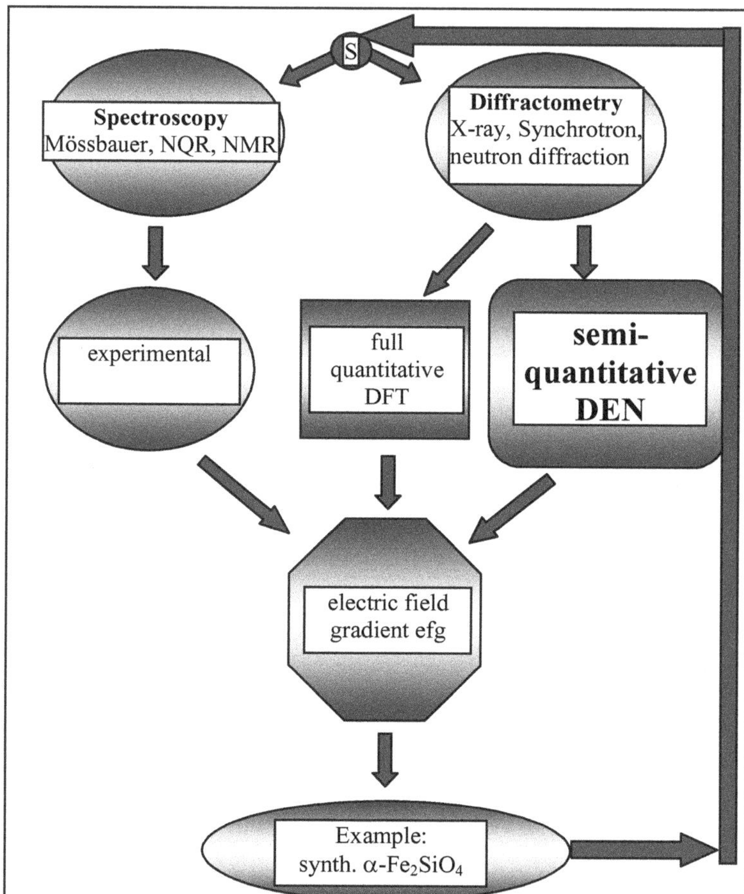

Figure 1.1 Functional scheme of the different methods involved at the evaluation of the electric field gradient efg as the common link of diffractometric, spectroscopic and density functional theory data.

Chapter 2

The Basic Quantity: The Electric Field Gradient (efg)

The efg is the common quantity of diffractometry and spectroscopy and can be determined experimentally as well as theoretically. As one might infer from the terminus, it is the first local derivative of the electric field and, hence, the second derivative of the electrostatic potential stemming from surrounding charges at the site of a given nucleus:

$$\text{efg} = \vec{\nabla}\vec{E} = -\vec{\nabla}\vec{\nabla}V = \begin{pmatrix} V_{xx} & V_{xy} & V_{xz} \\ V_{yx} & V_{yy} & V_{yz} \\ V_{zx} & V_{zy} & V_{zz} \end{pmatrix}, \tag{2.1}$$

$$\text{with} \quad V_{ij} = \partial^2 V/\partial i \partial j \ (i, j = x, y, z) \tag{2.2}$$

By diagonalisation of the above matrix it is possible to let the non-diagonal elements disappear in favour of the main diagonal elements V_{xx}, V_{yy} and V_{zz}. To this purpose different numerical methods have been described in the relevant literature, e.g. the Jacobi method (Stummel and Hainer, 1971).

The choice of the main diagonal elements is commonly performed according to the following hierarchy:

$$|V_{zz}| \geq |V_{yy}| \geq |V_{xx}| \tag{2.3}$$

The Difference Electron Nanoscope: Methods and Applications
Werner Lottermoser

ISBN 978-981-4774-01-7 (Hardcover), 978-1-315-19664-0 (eBook)
www.panstanford.com

Geometrically—and like many quantities in physics—the efg is represented by a tensor ellipsoid, what might be seen from the above matrix structure (Fig. 2.1).

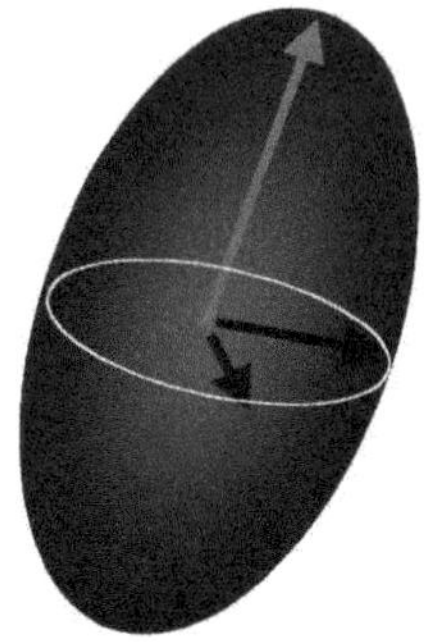

Figure 2.1 Graphical representation of the tensor of the electric field gradient in a quasi-3D plot.

Other examples from fundamental physics are the inertia and the refractive index tensor.

At this point another quantity should be considered, which provides the link between the efg tensor and the experimental methods for its evaluation: the asymmetry parameter η that describes the flattening of the efg tensor. It is defined as follows:

$$\eta = \frac{V_{xx} - V_{yy}}{V_{zz}} \tag{2.4}$$

This quantity is an experimental parameter that can be directly derived from Mössbauer spectroscopy or NMR/NQR (see Sections 3.1.1, 3.1.2 and 3.1.3). Together with the main component of the efg, V_{zz}, that is contained in another experimental parameter from Mössbauer spectroscopy or NMR/NQR, the quadrupole splitting ΔE_Q or QS, we have in favourite cases the means at hand to derive the main components of the efg, V_{xx}, V_{yy} and V_{zz}.

Which components we may find out in the concrete case, however, is dependent on the distribution of the surrounding charges that determine the shape of the efg.

In general, we have the following three possible cases (Figs. 2.2, 2.3, 2.4):

(i) The surrounding charge distribution with respect to the point of consideration (i.e. the position of the central nucleus) has a cubic point symmetry—this also includes regular tetrahedral or octahedral environments of charges (Fig. 2.2).

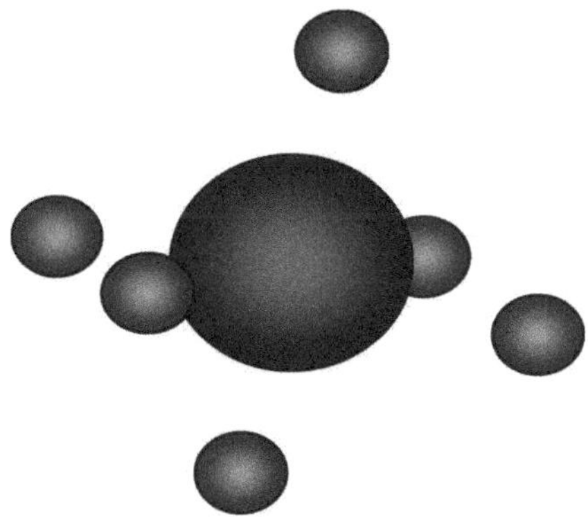

Figure 2.2 The efg tensor in a cubic environment of surrounding charges. In this case, the efg tensor ellipsoid is a sphere—the relevant axes cannot be derived: The main component of the efg, V_{zz}, and the asymmetry parameter η are zero.

(ii) The surrounding charge distribution with respect to the position of the central nucleus has a tetragonal point symmetry—this means a basically cubic environment which is distorted along one axis (Fig. 2.3):

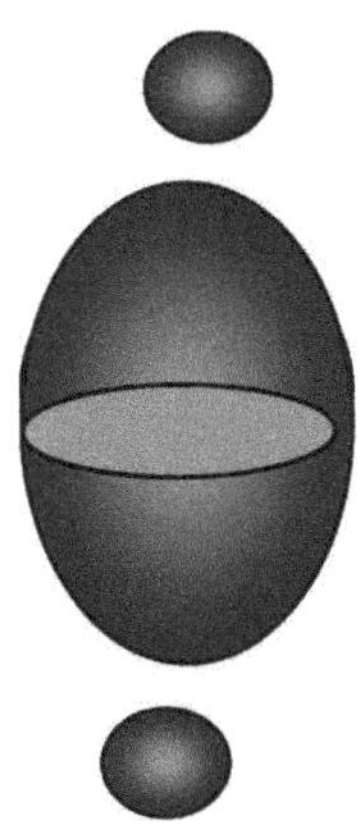

Figure 2.3 The efg tensor in a tetragonally distorted environment of surrounding charges. In this case, the efg tensor ellipsoid is rotational symmetric. The main component of the efg, V_{zz} can be derived ($V_{zz} \neq 0$) whereas the asymmetry parameter η is zero, since the components V_{xx} and V_{yy} are, due to the rotational symmetry, equal to each other.

(iii) The surrounding charge distribution with respect to the position of the central nucleus has a triclinic point symmetry—this means an arbitrary charge environment (Fig. 2.4):

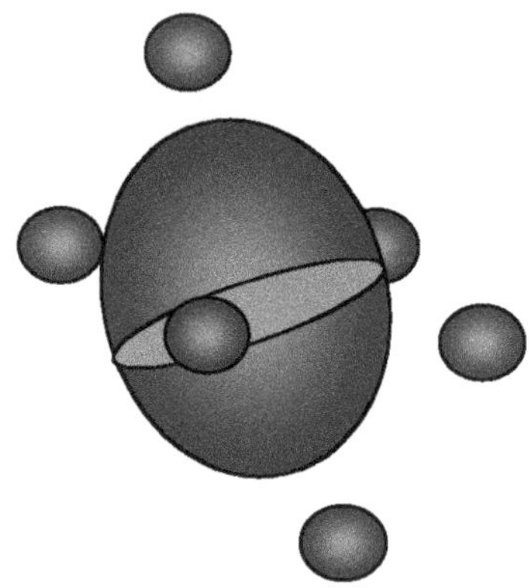

Figure 2.4 The efg tensor in a triclinicly distorted environment of surrounding charges. In this case, the efg tensor ellipsoid has rhombic symmetry. The main component of the efg, V_{zz} can be derived ($V_{zz} \neq 0$) as well as the asymmetry parameter η ($\eta \neq 0$), since all three components V_{xx}, V_{yy} and V_{zz} have finite values.

It remains to ask how such a distribution of surrounding charges might be determined.

In a first approximation and in the easiest case, where the charges are confined on very small locations as compared to the distances between them, the point charge model can be applied. Here, we get for the Coulomb potential of these charges in Cartesian coordinates:

$$V = \sum_i \frac{e_i}{r_i} = \sum_i \frac{e_i}{\sqrt{x_i^2 + y_i^2 + z_i^2}} \tag{2.5}$$

with e_i being the single charges and r_i the distances from the point of reference.

According to the formula 2.2, we can derive all the components of the efg from Eq. 2.5. However, this procedure would be very arduous to be performed in Cartesian coordinates. Since the problem is connected with nuclei and ions in a crystal, whose different sites have point symmetry, and the charges might be considered as a cloud with respect to a special site, it is convenient to switch to polar coordinates. Hence, the main component of the efg, V_{zz}, and the asymmetry parameter η might be written in the following notation:

$$V_{zz} = \sum_i e_i \frac{3\cos^2\theta_i - 1}{r_i^3}$$
$$\eta = 3/V_{zz} \sum_i e_i \sin^2\theta_i \cos 2\varphi_i / r_i^3, \tag{2.6}$$

with e_i being the single charges and r_i, θ_i and φ_i their polar coordinates.

In case of diffuse or largely overlapping charges, as it is intrinsic for the 3d ions treated in this book, the point charge model is not applicable in the easiest form. But we have other tools to overcome these difficulties (cf. Section 3.2).

Chapter 3

The Three Pillars of the DEN Method

3.1 The Experimental Methods to Derive a "Measured" efg

3.1.1 Fundamentals of Mössbauer Spectroscopy

Up to the 1950s and in spite of the preceding advances in the field of spectroscopy, it was commonly considered as being impossible to determine the energy levels and transitions within the ground and excited states of nuclei. This had mainly two reasons:

(i) Whereas transitions in the electronic shell of an atom can be easily measured by a common optical spectrometer due to the low lifetime of the excited state and, hence, the comparably broad line widths involved, this is not valid for the energy states of excited nuclei: Here, the lifetime of the excited state is comparably huge, the energy or frequency line width, on the other hand, comparably tiny. This effect is due to an uncertainty relation that may be written as

$$\Gamma = \frac{\hbar}{\tau}, \tag{3.1}$$

The Difference Electron Nanoscope: Methods and Applications
Werner Lottermoser

ISBN 978-981-4774-01-7 (Hardcover), 978-1-315-19664-0 (eBook)
www.panstanford.com

where Γ is the line width, τ the lifetime of the excited state and $\hbar$ Planck's constant.

The frequency region of the electromagnetic wave here involved is X-ray, whereas the spectra of excited electrons are in the range of the visible light. Switching to the particle image, we may consider the X-ray wave as a γ-photon. Figure 3.1 illustrates the physical properties that are applicable to quantum resonance phenomena in general.

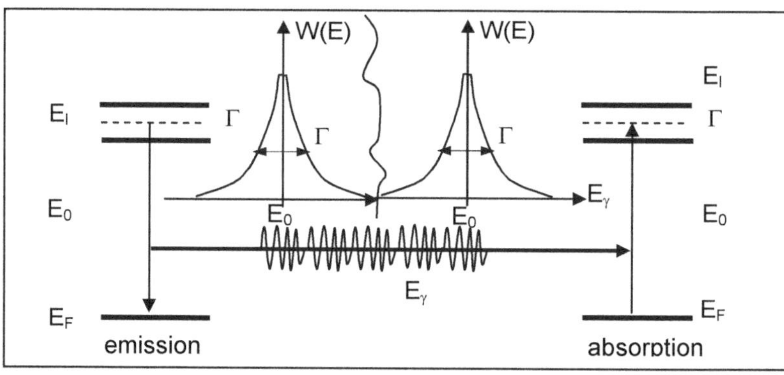

Figure 3.1 The energy levels of two neighbouring nuclei and the exchanged γ-photon.

Γ is the halfwidth of the lorentzian frequency line of the γ-photon, E_F the fundamental state, E_I the excited one and E_0 the transition energy. E_γ is described as an energy distribution $W(E)$ around E_0 that may be written as the famous Breit–Wigner formula:

$$W(E)=\frac{\Gamma^2/4}{(E-E_0)^2+\Gamma^2/4} \tag{3.2}$$

Due to the long lifetime of the excited energy state in the case of nuclei, the halfwidth of the lorentzian line of the γ-photon is incredibly small and uncompared to other effects in nature.

(ii) The second obstacle is the influence of the recoil. Whereas in the frequency range of electronic transitions in the shell and the subsequent emission of optical photons the hereby released recoil on the atom does not play any role; this is substantially otherwise in the case of excited nuclei: The influence of the recoil which decreases the energy of the emitted γ-photon is considerable. On the other hand, when such a photon is absorbed by an analogous nucleus, the recoil acts in the opposite direction, so that resonance at free nuclei cannot be observed. This is demonstrated by Fig. 3.2.

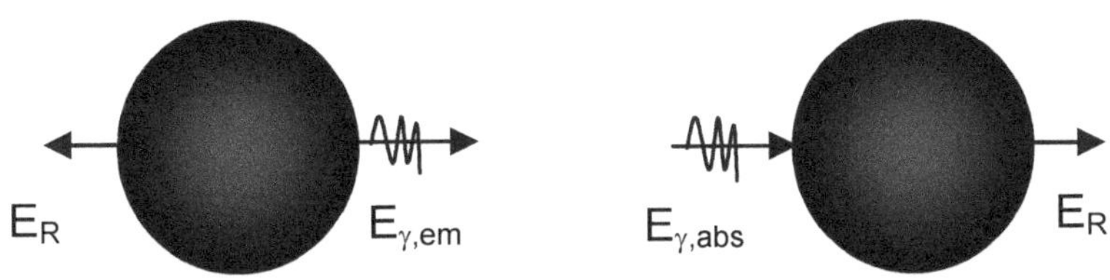

Figure 3.2 Emission and absorption of a γ-photon regarding the recoil involved, schematically.

The amount of the recoil can be estimated according to the equations mentioned below.

First of all, we may assume conservation of impulse and energy:

$$\vec{P}_\gamma + \vec{P}_{\text{nuc}} = 0 \qquad E_0 = E_\text{I} - E_\text{F} = E_{\gamma,\text{em}} + E_\text{R} \tag{3.3}$$

From this, we get for the energy of the recoil:

$$E_0 = \frac{P_{\text{nuc}}^2}{2M} \xrightarrow[\text{of impulse}]{\text{conservation}} \frac{P_\gamma^2}{2M} = \frac{E_{\gamma,\text{em}}^2}{2Mc^2} \tag{3.4}$$

With $E_0 = E_{\gamma,\text{em}} + E_\text{R}$ and $E_\text{R} << E_{\gamma,\text{em}}$ follows:

$$E_0 \approx E_{\gamma,\text{em}} \tag{3.5}$$

Using Eq. 3.5 we obtain:

$$E_\text{R} \approx \frac{E_0^2}{2Mc^2} \tag{3.6}$$

Hence, the energy of the emitted γ-quantum is reduced by the loss of the recoil:

$$E_{\gamma,\mathrm{em}} = E_0 - E_\mathrm{R} \approx E_0 - \frac{E_0^2}{2Mc^2} \tag{3.7}$$

The analogon is valid for the absorbed photon energy:

$$E_{\gamma,\mathrm{abs}} = E_0 + E_\mathrm{R} \tag{3.8}$$

Due to the recoil loss, the two energies of Eqs. 3.7 and 3.8 are not equal. Resonance effects are, however, only possible for overlapping amounts of the γ-emission and -absorption line, whereas the condition of resonance must be fulfilled:

$$2E_\mathrm{R} \leq \Gamma \tag{3.9}$$

This is illustrated by Fig. 3.3.

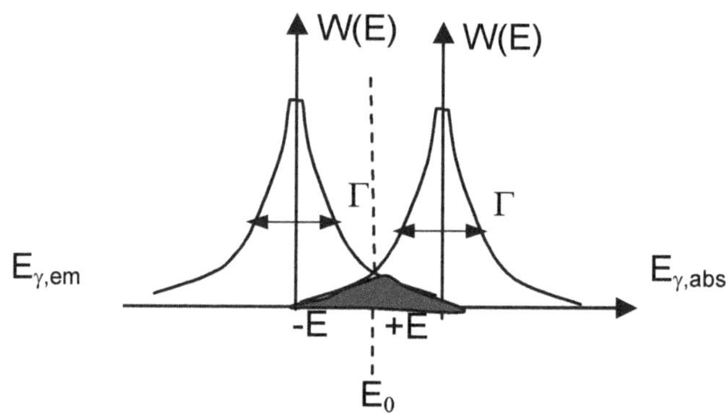

Figure 3.3 γ-emission and -absorption line with slight resonant amounts (grey coloured).

Whereas this resonance condition is always met in the optical case (due to the comparable low transition energy the recoil energy is also small), this is not true for transitions within the nucleus: The recoil energy is with respect to the line width Γ so huge that the resonance condition is commonly never met—resonance should be not observable in this case.

The reason for this may be demonstrated by using a convincing comparison:

A good measure for the sharpness of a resonance line is the ratio of the half-width at half maximum Γ and the energy E_0

of this line. In the case of a common receiver in the medium wave range this ratio amounts to ~0.01 (1:100). The frequency of the broadcasting station can be easily obtained by tuning the variable condenser in the receiver's resonance circuit. In the case of a common Raman Laser (in the optical range of the electro-magnetic spectrum) the above ratio amounts to $\sim 2 \times 10^{-6}$ (1:2 mio.). This is comparably low but can be handled nowadays without any problems.

In the case of Mössbauer spectroscopy, however, this ratio is only $\sim 0.3 \times 10^{-12}$ (1:0.3 billiards!). The distance between the two lines (emission and absorption) on the energy scale together with the incredibly low linewidth leads to the consequence that under normal conditions resonance effects with free nuclei are not observable.

However, there were efforts in the beginning of the 1950s to achieve resonance scattering of certain nuclei by putting them on ultra-centrifuges. By using orbital velocities of $\sim 10^5$ m/s it was possible to obtain some γ-resonance scattering (Moon, 1951).

However, the relevant lines were severely broadened by the Doppler effect, so that information about the nuclear energy levels could not be obtained.

The same was true for experiments with enhanced temperatures (Malmfors, 1955)—also in this case the procedure led to some resonance scattering but again at the expense of a reasonable low linewidth.

In this situation proved Rudolf L. Mössbauer, a young thesis student in the group of nuclear physicist Heinz Maier-Leibnitz, that resonance scattering of γ-rays of excited nuclei may also be possible by fixing emitting and absorbing nuclei inside of a crystal lattice thus eliminating the troublesome recoil. The upcoming resonance line had the "natural", very small line width Γ and—it was at the location of the fundamental energy E_0 on the energy scale. This discovery was achieved in the late 1950s by applying the Lamb theory of thermal neutron capture to a system comprising the nucleus bound in the lattice, the quantized oscillations of the lattice (phonons) and the electromagnetic radiation field (γ-photons). The solution of the Schrödinger equation with time-dependent perturbation theory yields an emission probability that is very similar to the above-mentioned

Breit–Wigner dependence $W(E)$, but with a more complicated structure. In general, three different cases may be distinguished[1]:

(1) The bonding of the nuclei to the crystal lattice is strong; this is equivalent to a small Doppler broadening of the resonance line compared to the Debye energy. The formula for $W(E)$ is analytically evaluable and one obtains a lorentzian peak at $E = E_0$ on the energy scale. This is equal to the case, where no energy at all ($0\hbar\omega$) is transferred to the lattice or—in other words: the recoilless amount of the γ-emission. This also corresponds to the "Mössbauer case" and is graphically represented in Fig. 3.4 (Amthauer and Pavicevic, 2001).

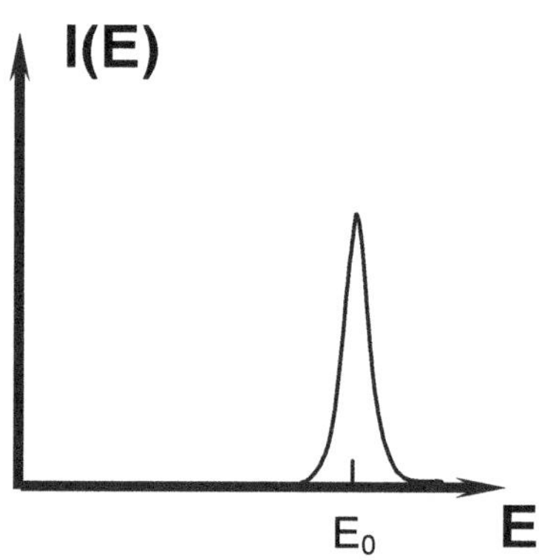

Figure 3.4 Energy distribution in resonant nuclear systems (complete recoilless—pure Mössbauer effect).

It is difficult to understand that a recoil should be completely quenched without changing the oscillation state of the lattice and, in classical physics, this is impossible indeed.

However, in terms of quantum mechanics, applied by R. L. Mössbauer this $E_R = 0\hbar\omega$ is explicitly included in the scenario.

The classical view often is an obstacle for many students to understand the Mössbauer effect. For a better and clearer

[1]The relevant measure for the strength of interaction in the occurring case is the ratio of the recoil energy E_R to the Debye-energy $k_B\Theta_D$ (Θ_D = Debye temperature) of the lattice. The latter is significant for the elastic properties of the crystal lattice in consideration.

viewpoint, imagine a cannon (the emitting system) on a ship. When firing the cannon, the energy will be distributed partly to the ship as the recoil (pushing the ship backwards) and partly to the cannonball providing the projectile's energy (this is analogous to the γ-ray energy). Now imagine the mass of the ship increasing more and more, up to infinity. When firing the cannon now, the recoil will be completely annihilated and the energy fully transferred to the projectile. It will also receive the complete expansion energy of the occurring gases (gunpowder) and a rather sharp energy distribution. This is, *cum grano salis*, the Mössbauer case in a rough analogous comparison.

(2) The bonding of the nuclei to the crystal lattice is weak; this is equivalent to a wide Doppler broadening compared to the Debye energy of the lattice. One obtains for the energy distribution $W(E)$ two ranges: The first one comprises the case of energy transfer of $E_R = 1\hbar\omega,\ 2\hbar\omega,\ 3\hbar\omega \ldots n\hbar\omega$; this leads to a broad Gaussian curve with a maximum at $E = \hbar\omega - E_R$. The second one again is equivalent to a lorentzian peak at $E = E_0$ (Mössbauer amount at $0\hbar\omega$).

The whole distribution may be viewed in Fig 3.5; this is the most occurring case: A pure "Mössbauer" amount with a simultaneous phonon excitation.

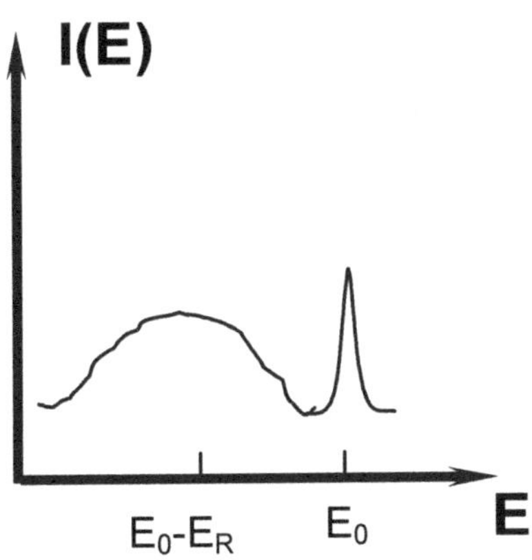

Figure 3.5 Energy distribution in resonant nuclear systems (partly recoilless—lattice contribution + Mössbauer effect).

(3) The nuclei are free or very weakly bound. The energy distribution in this case only comprises a very broad Gaussian hump—the recoilless fraction has disappeared completely. This case is realized in fluids or gases; hence, we observe no Mössbauer effect in these aggregate states (Fig. 3.6).

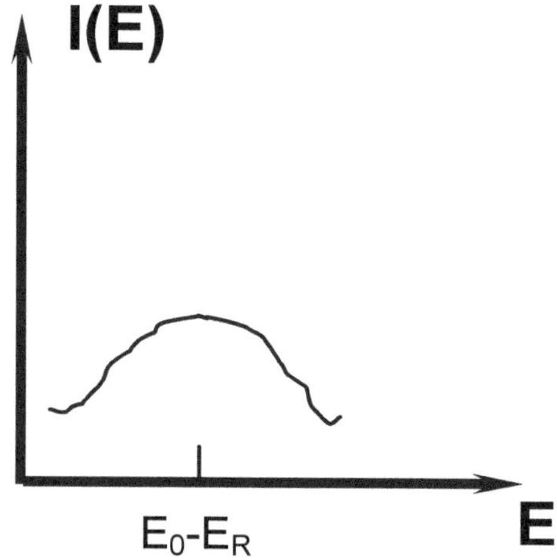

Figure 3.6 Energy distribution in resonant nuclear systems (only phonon scattering—Mössbauer effect disappeared).

A certain exception consists of the widely occurring glasses: Since a glass very often is described as a hypothermal fluid, the Mössbauer effect should not be observed here. In reality, most glasses are partly crystalline and partly amorphous; hence, we commonly observe a Mössbauer effect, but with very broad lines due to the non-defined lattice sites in the glass. This leads to a broad distribution of sites and hence, to a broadening of the line. Whether a so-conditioned line spectrum is evaluable or not, depends on the crystallinity of the glass and the intuition of the researcher.

In the whole, R. L. Mössbauer has become famous with his explanation of the effect named after him. More scientifically, the phenomenon is also called the "recoil-free emission and absorption of nuclear γ-rays". For his discovery R. L. Mössbauer received the Nobel prize in 1961 (at the age of 32!).

It is therefore so eminently important because from then on it has become possible to examine nuclear energy levels under most variable conditions.

Unfortunately, the effect does not occur within every nucleus that might be imaginable. Up to now, around 1500 "Mössbauer-active" nuclides have been detected, most of them being among the REE (rare earth elements) with predominantly very low half-lives. The latter is decisive for the practical applicability of a Mössbauer γ-ray source, since a Mössbauer experiment lasts about 3 days on average, and sources with a faster decay than this value are only of academic interest.

Fortunately, one of the most abundant elements on Earth, iron (^{56}Fe), contains an amount of up to 2% ^{57}Fe that is the Mössbauer active nuclide. This can be used as a probe or sensor on a special lattice site of the iron-containing compound. Only a tiny quantity of several milligrams of natural iron is normally sufficient to get an evaluable spectrum. The emission (or absorption) line, respectively, for ^{57}Fe arises from the 14.41 keV transition between the nuclear levels I = 3/2 and I = 1/2. These are angular momentum states of the ^{57}Fe nucleus with the excited state I = 3/2 and the ground state I = 1/2 (cf. Fig. 3.7, on the left).

The emitting source are ^{57}Co nuclei embedded in a comparably firm crystal lattice like Pd or Rh—through inner conversion they transform to the excited state I = 3/2 of ^{57}Fe which subsequently decays to its ground state with I = 1/2 within a half-life of 270 days emitting 14.41 keV γ-rays. This is more than sufficient to produce a Mössbauer source of comparably long lifetime to provide resonance absorption in ^{57}Fe nuclei contained in the "normal" ^{56}Fe of the chosen sample. Other nuclides with importance for different fields of applications are ^{119}Sn, ^{197}Au, ^{121}Sb, ^{125}Te and ^{151}Eu.

Up to now, we have mentioned only one Mössbauer line due to one transition between the nuclear energy levels. However, the line may be, corresponding to the levels, shifted against the zero point E_0 of the energy scale and/or split off by different electric or magnetic interactions for the ^{57}Fe nucleus with its surroundings in the crystal lattice. These are the so-called hyperfine interactions.

Within the electric interactions we distinguish between two main contributions:

(1) The relevant potential is due to charges within the nucleus (electric monopole term).

This contribution results from the Coulomb interaction between the positive charge of the nucleus and the electron density $|\Psi_s(0)|^2$ at the site of the nucleus.

It may be surprising to some readers that electronic charges (from the shell) might emerge within the nucleus—contrary to every classical imagination. The reason is that, within the frame of quantum mechanics, we have to consider electrons not circulating on orbits but remaining on hyperareas with maximum probability, the spherical symmetric s-orbitals in the ground state. According to this image, the electrons do either not circulate on the surface of these orbitals, but perform a sort of rosette or pendulum course with maximum local probability at the reversal points (the s-orbital). However, in between exists a small but non-vanishing probability for these s-electrons to emerge into the nucleus thus transferring "information" about oxidation (valence) or spin states to the nucleus. The result is a shift of the undisturbed nuclear energy levels and thus of the single Mössbauer line from the site E_0—this shift is different for nuclei in different energy states (isomers) and is therefore called "isomer shift δ". Its value is characteristic for the oxidation state of the ion (e.g. whether we have to deal with Fe^{2+}, Fe^{3+} or other). Due to screening effects the s-electron charge density is changed by different occupation of d- or f-orbitals, so that, e.g., the six 3d-electrons of Fe^{2+} screen better (higher δ-value) than the five 3d-electrons of Fe^{3+} (lower δ-value).

Through the chemical bonding we also might get information about different coordinations (tetrahedral, octahedral, a.s.o.) around the given ion/nucleus, which may be deduced from the δ-values as well.

(2) The relevant potential is due to charges at the outside of the nucleus (electric quadrupole term).

This contribution results from the interaction between the quadrupole moment Q of the nucleus and the electric field gradient (efg) V_{ij} and leads to a splitting of the excited nuclear energy level—the resulting two transitions yield instead of the former single line a doublet of two lines.

The distance of this pair on the energy scale is called the quadrupole splitting ΔE_Q and is directly related to the above-mentioned efg tensor (cf. Chapter 2).

$$\Delta E_Q = \frac{1}{2} eQV_{zz}\sqrt{1 + \eta^2/3}, \tag{3.10}$$

where e is the electron charge unit, Q the nuclear quadrupole moment, $V_{zz} = eq$ the main component of the efg and η the asymmetry parameter (cf. Chapter 2) which describes the flattening of the efg tensor ellipsoid. The shape of the latter under the surrounding of different symmetries has already been described in Chapter 2.

ΔE_Q is only non-vanishing at primarily aspherical electronic distributions around the given nucleus with two main contributions:

(1) Through ΔE_Q of the Mössbauer lines, it is possible to get information about unfilled or distorted electronic shells, e.g. p- and d-orbitals. The degeneration of the energy levels is partly or fully raised, so that ΔE_Q also contains the influence of the surrounding crystal field (Fig. 3.7).

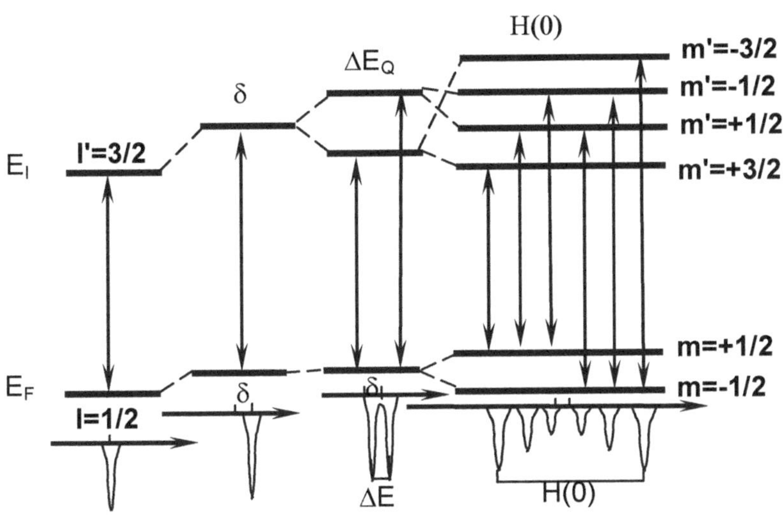

Figure 3.7 Possible shifts or splittings of the ^{57}Fe hyperfine levels and the relevant transitions (double arrows). δ = isomer shift, ΔE_Q = quadrupole splitting, $H(0)$ = internal magnetic field leading to magnetic hyperfine splitting, E_F = fundamental nuclear state with $I = 1/2$, E_I = excited nuclear state with $I' = 3/2$.

From the experimental quadrupole splitting it is therefore possible to deduce the surrounding charge distributions of the given nucleus—this is essential when determining the probable lattice site of the ion/nucleus under consideration. For example, octahedrally coordinated sites provide other values of ΔE_Q than those with a tetrahedral environment. Since every non-equivalent crystallographic site provides one Mössbauer subspectrum, the superimposed total pattern may often be only distinguished with respect to the single contribution by evaluation of ΔE_Q and η (cf. the example of synthetic fayalite, Chapter 5).

(2) The lattice contribution to the efg is directly provided by Coulomb interaction with the surrounding charges outside of the electronic shell (ligands). In general, this lattice contribution is smaller than the one of the valence shell. However, the relative influence of both on the efg is balanced by the Sternheimer screening and anti-screening factors (Sternheimer, 1963), so that the total effect is not easily predictable.

For example, in the case of high-spin Fe^{3+}, five 3d-electrons occupy all five orbitals with quasi-spherical charge distribution around the nucleus. So only the comparably low lattice amount will provide a contribution to the total efg. In the case of high-spin Fe^{2+}, however, six 3d-electrons are distributed over five 3d orbitals with one spin being antiparallel to the others. This causes such a strong asymmetric charge distribution in the Fe^{2+} shell that the valence contribution will be dominant. By this, it is evident to expect a stronger quadrupole splitting for Fe^{2+} than for Fe^{3+}—this is experimentally confirmed.

(3) Another important physical property, also with respect to the DEN, is the magnetic dipole interaction between an internal magnetic field $H(0)$ at the site of the nucleus and its dipolar moments. This also leads to a splitting of the nuclear energy levels, the excited one in four sublevels (characterized by the magnetic quantum number $m' = \pm 3/2, \pm 1/2$) and the ground state in two sublevels with $m = \pm 1/2$. This causes together with the selection rules $\Delta m = \pm 1,0$ six allowed transitions with a six-fold splitting of the Mössbauer line into a sextet (Fig. 3.7).

The description of the magnetic interactions yields several possible contributions to $H(0)$:

(1) External magnetic fields—it is possible to apply a magnetic field to a given sample—and their effects within the sample (H_{loc}).

(2) The magnetic fields of electronic origin. Main contribution here is the so-called Fermi-contact term which arises from non-compensated spin densities from s-electrons at the site of the nucleus (due to completed and non-completed electron shells). Paired s-electrons can be polarized by unpaired 3d-electrons—an effect of a combined exchange-/Coulomb interaction.[2]

By this, it is possible to explain the strong negative fields $H(0)$ at the site of Fe^{2+} and Fe^{3+}. Hence, the d-electrons will exert a strong influence on $H(0)$—however, conclusions leading too far should be avoided, since a large magnetic moment not necessarily yields a large internal magnetic field!

However, not only $H(0)$, e.g. in Tesla units can be derived, it is also possible to determine its orientation with respect to the axes of the efg through two angle parameters θ and ϕ which might also be derived by a Mössbauer spectrum refinement procedure.

The combined shift and possible splittings of the nuclear energy levels for the different interactions are summarized in Fig. 3.7 above. The different transitions together with the relevant line intensities are denoted as arrows or lorentzians, respectively.

3.1.1.1 Features and function of a Mössbauer spectrometer

Up to now, we have not yet described a practical method to record a Mössbauer spectrum. Since R. L. Mössbauer's discovery several types of spectrometers for the practical application of the effect have been developed. The basic principle of all those, however, is merely the same:

The intensity of the transmitted or scattered γ-rays is recorded dependent on the relative velocity between source ("emitter") and absorber ("the sample"). By the previously described phenomena (isomer shift, quadrupole and magnetic

[2]The exchange interaction is responsible for collective magnetic phenomena, e.g. ferromagnetism, and has first been postulated by Werner Heisenberg.

hyperfine interactions), the position of the line and its splitting will be changed with respect to the undisturbed line—the condition for resonance will be lifted. By changing the relative velocity between source and absorber (i.e. exploiting the linear Doppler effect[3]) the frequency and thus the energy on the horizontal scale will be varied until again resonance is observed. The velocity can be varied so exactly that even the profile of the extremely narrow Mössbauer line can be determined. This is in particular important in those cases where several subspectra (due to non-equivalent crystallographic sites of the nuclei or different crystal phases) are superimposed. The whole procedure is comparable with the automatic tuning in a modern radio receiver (claiming this previous analogous comparison again): The frequency (or wavelength, respectively) of the receiver is changed in constant or accelerated steps till the emitting frequency of the radio station is met. We will then observe maximum audibility of the transmitter—in this case we may say that transmitter ("the station") and radio are in resonance. Analogously, the radioactive source (^{57}Co with decay to ^{57}Fe excited state and subsequent decay to ^{57}Fe ground state under emission of Mössbauer γ-rays) will be at special relative velocities of the source—and hence of the γ-photons—in resonance with the ^{57}Fe nuclei of the absorber ("the sample"). Since the intensity is transmitted through the absorber, we will get a decrease of the measured resonant absorption at these special velocities. Therefore a Mössbauer spectrum is commonly turned upside down (with the resonance peaks displayed as minima). The resonant absorbed intensity dependent on the momentary relative velocity is recorded as Mössbauer spectrum. The velocity values at these resonance peaks are now highly evident for the values of the Mössbauer parameters isomer shift, quadrupole splitting, line width a.s.o. (see above).

The great advantage of (Mössbauer) spectroscopy is that these parameters can roughly be gained from a rather quick inspection of a spectrum plot by help of a simple ruler (contrary to, e.g., diffractometry). For more sophisticated evaluations, however, we need a refinement program (see below).

[3]The linear Doppler effect is primarily the variation of the sound frequency by variation of the relative velocity between source and observer of an acoustic wave. This effect is one of the most principal and important phenomena in modern physics.

The determination of well-resolved spectra is a severe challenge for the applied spectrometer. In many labs, a special device with "constant acceleration" is in common usage—a schematic view is displayed in Fig. 3.8.

The driver G (a high-accuracy function generator) generates a signal g (constant acceleration = linearly tunable velocity) acting on the spindle of a linear motor T—in fact a highly sophisticated loudspeaker—with the source Q on top of the spindle.

Q is in most cases a small screw, in whose head a platelet with the radioactive substance (here: ^{57}Co in a Rh lattice) is fixed and covered by an inert sheet of acrylum. The screw is attached to the spindle and moved back and forth with linearly varying velocity.

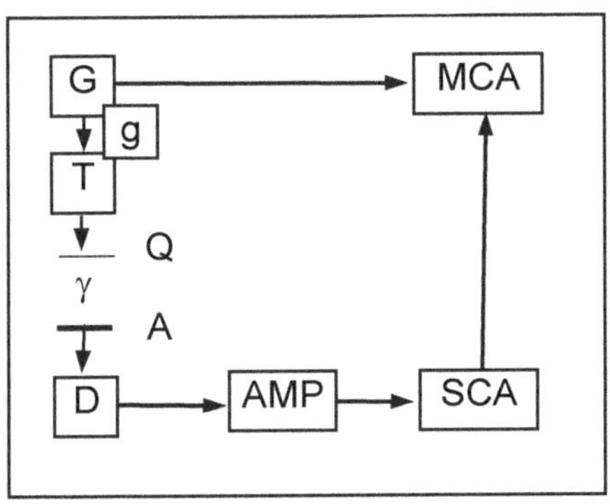

Figure 3.8 Mössbauer device, time-mode *T*.

The initially analogous signal g (Fig. 3.9) is digitized—each velocity value is attributed to a velocity channel and each count rate per velocity channel is transmitted to the multichannel analyser MCA. Here, all the different count rates arriving singularly from the SCA are stored subsequently in 1024 (2^{10}) velocity channels. It is evident that there must be a connection from G to the MCA in order to synchronize the MCA with the SCA. The result is a counting rate (absorbed intensity) dependent on the velocity channel, which is subsequently read out as Mössbauer spectrum.

The equally driven γ-photons move to the absorber A (the sample) and release there resonant absorption in the ^{57}Fe nuclei of the sample at certain velocities. In the case of non-absorption (non-resonance) the γ-photons are transmitted and form the background in the detector D (commonly a sort of Geiger-Müller counter tube). At special energy/frequency or velocity values, respectively, we observe a sudden decrease of the transmitted intensity—the resonantly absorbed Mössbauer peaks. The count rates (= intensities) are amplified (AMP) and stored in a single channel analyser (SCA) for each velocity value.

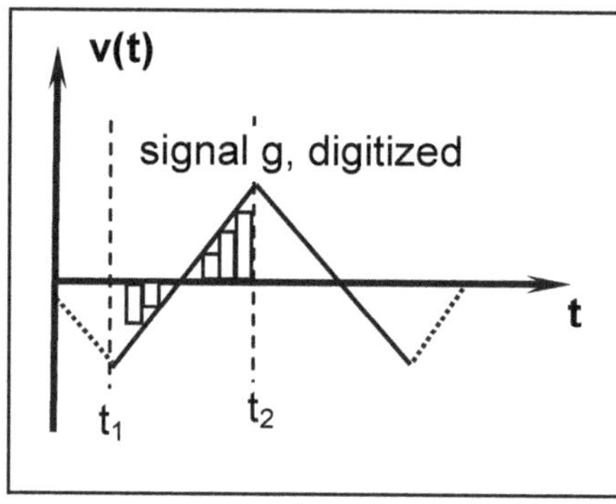

Figure 3.9 Velocity slope of motor configuration. For explanation please see text.

3.1.1.2 Evaluation of spectra

In particular for combined hyperfine interactions (isomer shift + quadrupole splitting + magnetic hyperfine splitting) and for several crystallographically non-equivalent sites yielding several superimposed subspectra the resulting spectrum is so complex that it can only be solved numerically by help of computer programs.

The basic principle of most of these routines is rather uniform: It consists of the numerical solution of the eigenvalue equation of the Schrödinger formalism. First of all, the matrix elements (parametrized by the o.a. electric and magnetic quantities) of the total Hamiltonian H_{hfs} for all eigenstates $<I,\ m|\ H_{\text{hfs}}|\ I,\ m'>$ are set up. By diagonalization, e.g. according to the Jacobi procedure, the secular equation of the energy eigenvalues will subsequently be solved. By this, the line positions are determined, which may be transformed to values of relative velocities. For the calculated energy levels the transition probabilities and thus the relevant intensities are calculated. The result is a theoretical Mössbauer spectrum with Y^0_{cal} $(x_i,\ \pi^0_1\ \ldots\ \pi^0_1)$—the Y_{cal}

are the calculated intensities, x_i the channel numbers and π_i the Mössbauer parameters to be refined (isomer shift, linewidth, a.s.o.).

In general, this theoretical pattern will deviate from the observed spectrum Y_{obs}. Applying the method of the Least Squares (Gauss-procedure), the above-mentioned spectral function around the initial parameter values π^0 will be developed and an improved parameter set π^1 calculated. This is used as new initial values with an improved refinement of the observed pattern. This iteration procedure will be continued, until the function (goodness-of-fit parameter):

$$\chi^2 = \sum_{i=1}^{N} \left[Y_{cal}(x_i, \pi_1 \dots \pi_n) - Y_{obs}\right]^2 w(x_i)/(N-k) \quad (3.11)$$

adopts a minimum. N is the total channel number, $w(x_i) = 1/Y_{obs,i}$ the weighing scheme and k is the number of refined parameters.

The Mössbauer parameters for the calculation of a well-refined theoretical pattern together with their units are the following:

δ (isomer shift, mm/s), Γ (half-width at half maximum, mm/s), I (netto intensity, counts), QS (quadrupole splitting, mm/s), $H(0)$ (internal magnetic field, T), asymmetry parameter η, collateral angle θ and azimuthal angle φ[4].

In most cases, Mössbauer spectroscopy is applied to powder samples—for the preparation of the relevant absorbers, see Section 3.2. Here, in the powder case, the orientation of the efg with respect to the crystallographic axes can only be deduced indirectly and with high uncertainty because it is merely based on symmetry assumptions.

3.1.1.3 Calibration and folding

In order to compare the Mössbauer spectra of different spectrometers, it is necessary to perform a calibration of the abscissa from channel numbers (output of the MCA) to absolute values of the relative velocity between source and absorber. For this purpose, a standard substance is measured before and/or after recording the Mössbauer spectrum of the desired sample with identical maximum velocity setting. The Mössbauer effect is so accurate that even slight tilts at the wire potentiometer of

[4]The direction of the internal magnetic field with respect to the axes of the efg and the relevant orientational angles θ and φ are shown in Fig. 3.10.

the driver unit are sufficient to obtain quite another relative velocity and thus, other line positions. Hence, it is not possible to find identical adjustments of the velocity only by inspection of the digital display of the driver unit—this is much too inaccurate. However, nowadays there exist expensive driver units with a calibration based on optical interferometry—here, absolute velocity calibrations are possible *in situ*. In the case of most common spectrometers, the above-mentioned calibration spectra of a standard substance, e.g. an iron foil, have to be recorded. The resulting pattern contains two six-line magnetic hyperfine spectra connected by a mirror plane between each of them. This is due to the fact that the driver signal g (see Fig. 3.8) consists of a symmetrical sawtooth in most cases, with a linear increase in relative velocity up to a maximum value (to be adjusted at the driver unit) and a subsequent decrease. This sawtooth is repeated periodically and leads, due to its two mirror symmetrical branches, to two recorded Mössbauer spectra.

In order to make the second one identical to the first, the second spectrum has to be reversed by a computer program thus getting also a positive velocity slope. Now the two spectra can be directly compared. The distances of special lines in channel numbers are now determined by a computer graphical representation of the spectra and correlated with distances in mm/s from an absolute velocity measurement at the standard substance found in special tables. Hence, a calibration factor can be calculated which transforms the line positions from channel numbers in mm/s, the unit of the relative velocity.

This calibration factor is valid for all subsequent or precedent measurements with identical maximum velocity. By comparison of the line positions of the "left" and "right" Mössbauer spectrum it can be detected, e.g., whether the sawtooth is symmetrical, whether there are deviations in linearity, whether there are parasitic oscillations a.s.o. If necessary, the measurement has to be repeated.

The values of the count rates are easier to manage: Here, the minimum count rate (maximum absorption) with respect to the average background values (0% absorption = 100% transmission) can be transformed in ordinate values of "relative absorption in %", a common measure in many Mössbauer publications.

Having detected by calibration that everything is o.k., the two measured Mössbauer spectra of the desired sample may be "folded". This is no folding in the mathematical sense, but it is an averaging of the corresponding line intensities left/right by determining the position of the mirror plane, thus improving the count statistics by the factor 2.

3.1.1.4 Sample preparation (powder)

Last, but not least, a few words to the sample preparation of a Mössbauer powder absorber should be left. Many readers might it consider as trivial, but it should be emphasized nevertheless: Since Mössbauer spectroscopy is extremely sensitive to, e.g., iron-containing specimen, the usage of iron or steel mortars and friction elements in order to grind bigger crystal pieces is completely out of question! Automatic oscillation mortars are, even when consisting of iron free materials like agate, not to be used either, since they tend to produce a frictional heating and thus, an artificial enhancement of the Fe(III) contents.

Grinding of most samples should be performed manually in agate mortars with agate rods using acetone in order to reduce "sparkling"—therefore glasses should be worn during the procedure. In case of sheet silicates and other "smearing" substances, tempering with liquid nitrogen is recommended in order to make the sample brittle. The powder should finally be as fine as to be used for X-ray diffraction purposes i.e. no crystal faces should be visible any more.

Concerning the mass of the sample, a rule of thumb claims that 5 mg pure iron per cm^2 of absorber area is a good compromise between narrow linewidth and well detectable absorption. Ideally, the Mössbauer absorber should consist of 1 atomic layer, but in this case the intensity would be too low.

There are studies specifying the rule of thumb for concrete cases—if possible, several absorbers with different iron contents should be measured in order to determine the optimum empirically.

The absorber itself is prepared by using a copper ring of about 1 cm inner and 3 cm outer diameter and 0.1 cm thickness ("donut") with a thin sheet of super pure aluminium glued behind the hole—normal Al must not be used because of iron traces in the Al. The aluminium serves as a suppressor of higher excitation lines and for better heat conductibility. In order to avoid texture effects, the powder sample is commonly mixed up homogeneously with a filler material, say, dust sugar and subsequently pressed into the ring hole. However, this is only recommended for room temperature experiments and if the sample should be regained. In all other cases, in particular for cooling or heating experiments, the sample should be mixed up with a liquid, slowly hardening epoxy glue (e.g. araldite™). The sample can thus be made almost ideally homogeneous and remains inert for years after hardening.

The material of the ring should be chosen dependent on the application. For temperature dependent experiments copper provides a very good heat conductibility, however, the incoherent background will be enhanced. Therefore, for exclusive measurements at ambient temperature plastic rings are a good choice. However, from own experiences it may be stated, that very often further experiments at low or high temperatures turn out to become necessary, so that the "all-round" material should be applied from the beginning.

3.1.1.5 Historical background

The expression "spectroscopy" from the Latin word "spectrum" and Greek "scopein" (to observe) is a comparably old branch of natural science and originates from Isaac Newton who at first described the fraction of visible light into its different colours by transmission through a prism in 1704. Whereas this had merely been the observation of the pure phenomenon, spectroscopy as a physical tool for sophisticated experiments was established by Kirchhoff and Bunsen in 1859. Hence, we may define spectroscopy as a group of physical methods by which a radiation is analysed according to energy, wavelength, mass a.s.o. The distribution of intensity is called "spectrum" and the recording tools are named as spectrometers by which the link from pure observation (spectroscope) to quantitative measurement (spectrometer) has been performed. The examined radiation comprises the whole range of electromagnetic and mechanical waves, as well as non-classical "waves" like electrons, ions, atoms or molecules.

Three properties may be examined: The one of the radiation itself, the one of the source of radiation (emission spectroscopy) or that of a medium between source and detector (absorption/transmission spectroscopy). We distinguish between continuous and line spectra (sometimes combined) whereas for purposes of an exact determination of a wavelength a line spectrum is to be favoured.

Whereas the spectrum of a "black body" is continuous, the spectrum of excited "hot" gases shows coloured emission lines which, on the background of a continuous spectrum, may be recorded in absorption as black lines in order to determine the chemical elements of the absorbing sample. This was nicely demonstrated by the Fraunhofer lines in the spectrum of the sun, where the attributed elements are contained in the corona. Further on, in the end of the 19th and the course of the 20th century, the spectroscopic observations of atomic and molecular line spectra provided decisive hints for the development of modern atomic physics and quantum mechanics. It should be mentioned at this point that the very exact determination of fractions of wavelengths of the relevant spectrum lines enabled us to define fundamental physical quantities (like, e.g., the length of the unit meter). How this has been achieved?

Whereas in the early days of spectroscopy, prism spectrometers were widely used, very highly resolved lattice patterns have been applied. The latter act on the basis of interference (not of fraction like the prisms) and allow to detect a wavelength with high accuracy. Neighbouring "scratches" on a suitable target scatter an incoming light beam, analogously to the diffraction at a slit, and produce interfering waves. The sharpness of the diffracted maxima increases with the number of "scratches" (Airy sharpness). Since it is possible to produce several thousand "scratches" per centimetre, the wavelength resolution will become very high. Commonly, a spectrum comprises an intensity (or power density) dependent on energy or wave number. The energy on an atomic scale may be described by Planck's formula:

$$E = h\nu, \text{ with Planck's constant } h \tag{3.12}$$

The basis for the understanding of atomic spectra is the model of Bohr (however, nowadays, the orbital model is commonly acknowledged), where the electrons surround the atomic nucleus on certain orbits. The transition of an electron to a higher orbit leads to the absorption of a photon, whereas the back-drop to the lower orbit produces an emission with the photon energy:

$$\Delta E = E_i - E_f = h\nu \text{ (i = initial state, f = final state)} \tag{3.13}$$

in analogy to Section 3.1.1.

The concrete spectrum is also dependent on number of atoms, transition rules and occupation numbers of the different atomic levels.

Contrary to excited nuclei, however, the recoil plays no role in the case of photon emission/absorption in the atomic shell—the absorption of such a photon can easily lead to an electron transition to a higher shell, subsequent backdrop and again emission of a photon. This also may be interpreted as a resonance effect. Therefore it was not quite evident for the researchers of the 1940s that the spectra of γ-rays of free nuclei like in gases or fluids do not show any resonance effects. The reason as already mentioned above (Section 3.1.1) is the extraordinarily high lifetime of the excited nuclear levels; according to the uncertainty relation:

$$\Gamma = \hbar/\tau.$$

The half-width of the relevant spectrum line is extremely narrow and cannot be observed under "normal" conditions. Moreover, as mentioned above in Section 3.1.1, the recoil of the involved nuclei leads to a wide separation of the very narrow emission and absorption lines on the frequency scale, so that resonance can "normally" not be achieved. Several trials have been performed in the beginning of the 1950s by Moon (1951) and Malmfors (1955) (cf. Section 3.1.1) by artificially broadening of the resonance lines but the main advantage of the γ-lines, the possibility of measuring nuclear energy levels, has been sacrificed.

Then, in 1957, R. L. Mössbauer showed the famous way out of the dilemma by fixing the involved nuclei in crystal lattices hence erasing the fatal recoil and establishing one of the most sophisticated spectroscopic methods in modern physics (Section 3.1.1).

3.1.1.6 The Nobel Prize winner Rudolf L. Mössbauer

A book with a method that is widely based on the famous effect discussed above, should also demonstrate respect for the here-connected scientist himself. The author has had the great honour and pleasure of knowing him personally, so it is possible to enrich the commonly cited biographical data by some personal remarks[5].

[5]Standard biographical details of this and other famous personalities in this book have been taken from www.nobelprize.org.

Rudolf Ludwig Mössbauer was born on January 31, 1929, in Munich. After his matura in 1948 he was studying physics (highly motivated by the famous experiments at the Deutsche Museum at Munich, an outstanding exhibition of physical experiments and technical devices) and received diploma in 1955. The subsequent thesis work with the topic "Nuclear Resonance Fluorescence of γ-rays in ^{191}Iridium", in whose course he detected "his" effect, was not without friction. As he told me much later, his mentor Heinz Maier-Leibnitz wanted him to search for the effect at high sample temperatures, because HML thought to find it under these circumstances. Mössbauer, on the contrary, was sure—from theoretical reasons—to detect the effect at low temperatures. So he went to the Max-Planck-Institute of Medical Research at Heidelberg. There, he detected "his" effect as he had predicted—becoming more intense at low temperatures due to an enhanced f-factor that describes the effectiveness of the resonance line. In those days, experiments of this kind were very arduous—RLM had to construct most of his experimental equipment by himself including electronic components. There were almost no standard devices available, which are at our disposal nowadays; even the gear to produce the relative velocity between source and absorber was a sort of "lighthouse" driver with a rotating source, not with a linear motor ("loudspeaker"). As RLM told me, either the idea of using the Doppler effect via the relative velocity between source and absorber in order to tune the frequency/energy of the incoming γ-quants was born during a relaxing phase in his office. This idea enabled him to find the effect *experimentally* whereas a lot of theoretical work (highly sophisticated quantum mechanics) served him to predict the effect correctly. So he was the only physicist (at least whom I am knowing of) who was equally excellent in experimental and theoretical physics and—last but not least—as an academic teacher.

However, which were the events around the winning of the Nobel Prize?

As he told me, a special "messenger" from Stockholm visited him and wanted to know every detail of RLM's thesis work. Thus it came out that he—fortunately—did not obey the original directives to measure at high temperatures, but made his experiments at low ones as correctly predicted by him. Hence, it was decided that RLM should receive the Nobel Prize for the detection of the effect alone—as a young doctor of 32 years! In fact, he received it together with Robert Hofstadter in 1961, but the latter got it for other merits. Unlike many colleagues, R. L. Mössbauer returned to Munich after having spent several years at famous CalTech, USA. The Technical University at Munich was wise enough to obey his wishes concerning the introduction of a department structure for the scientific organization in order to enhance the scientific output. From 1972 till 1977 he was one of the directors of the ILL (Institute Laue-Langevin) at Grenoble and devoted his further activities to the detection of the neutrino rest mass. Experimental difficulties—the noise of the involved bolometers could not be suppressed sufficiently—however, hindered his ambitions for a second Nobel Prize. After his retirement he turned towards a third big topic in physics—cosmology.

As a sort of legacy he asked me, always to follow my own scientific way, as he had done before so successfully. His impressing perseverance helped him to maintain his (correct) predictions about "his" effect and also, to keep up against all obstacles. One striking example: Once he was in a talk show together with several high state officers. The latter stressed the importance of always knowing the most recent prescriptions and laws published in the German "Ministerialblatt" and asked RLM whether he was doing so. He just smiled and told them that, due to his giant pile of important work, he would just throw it into the waste paper basket when receiving them. There was big protest from the officers after that statement—but what could they do?

Like many of his famous colleagues, RLM always remained admiringly modest (in spite of his around 13 honorary doctor titles!) and was a wonderful narrator and teacher. In 2003, I invited him to give a speech at a symposium at Salzburg concerning the Doppler and related effects. In our spare time and on the way to his hotel in the nightly streets of Salzburg we were discussing a lot and I was impressed by his good physical condition in the age of 74! He told me that he has climbed on all important Bavarian mountains.

Unfortunately, his good health did not persist very long—he died on September 14, 2011 at the age of 83; many colleagues and I will never forget this splendid scientist and wonderful human being.

3.1.1.7 Later contributions to...

(a) General Relativity Theory

In spite of its confinement to around 45 elements and 1500 nuclides, the Mössbauer effect has a lot of applications in many fields of natural sciences. Over 1000 publications per year were at the culminating point of the "hype" not unusual. One of the first was the detection of the gravitational red shift of γ-rays by Pound and Rebka (1960). For this famous experiment to prove the predictions of the General Relativity Theory of Einstein, the extraordinary small linewidth of the nuclear γ-radiation has been utilized. Via Doppler effect, it is possible to tune the occurring frequencies and thus the resonances by changing the relative velocity between source and absorber. On the contrary, when the line positions of source and absorber are equal and no relative velocity between both present, there is no resonant absorption. Compensating for the temperature effect (low temperatures are chosen in order to enhance the *f*-factor) the source or the absorber, respectively, are brought to such a definite velocity that resonance absorption is measured. Now very small changes in frequency due to the tiny line width can be determined: From the instantaneous velocity of the source during absorption the frequency change of the γ-photon due to "acceleration" or "braking" can be measured.

For this purpose, the source and the absorber were mounted at a vertical distance of 22.56 m. During the first experiments the γ-quants from the source at the bottom moved opposite to the gravitational field and received an energy loss (red shift of the frequency $\Delta\nu_{grav} < 0$). During the opposite experiment the γ-quants moved from the source at the top of the "tower" parallel to the gravitational field and received an energy gain (blue shift of the frequency $\Delta\nu_{grav} > 0$). The difference of the shifts led to twice the effect for the single distance. Hence, it corresponds to a pure red shift experiment or a pure blue shift experiment with a distance of 45 m. The result was a shift of -4.2 ± 1.1 10^{-15} $\Delta\nu/\nu$ (prediction: -4.92 10^{-15}) and, with better statistics, -5.13 ± 0.51 10^{-15}. This corresponds with an accuracy of 10% to the value of the prediction and is therefore a splendid prove for the GRT and the accuracy of the Mössbauer method, respectively.

(b) The Gol'danskii-Karyagin Effect

During the calculation of the intensity of the Mössbauer lines due to a multitude of nuclei, the amount of the recoil-free events has to be considered. As mentioned above, is the latter described by the *f*-factor:

$$f = e^{-\langle(\vec{k}\,\vec{u})^2\rangle}, \tag{3.14}$$

where **k** is the wave vector of the γ-rays and **u** the shift from the local equilibrium in the lattice.

If the mean squared displacement value <(**ku**)^2> in the direction of **k** is equal for each propagation direction of the radiation, the *f*-factor will be constant. In some cases, however, it is possible that this factor will depend on the direction of the γ-rays. Then the calculated intensities will also depend on the directions $k(\beta,\gamma)$:

$$\overline{I}_o = \int_\gamma \int_\beta f(\beta \cdot \gamma)\, I_o(\beta,\gamma) \sin\beta\, d\beta\, d\gamma \tag{3.15}$$

instead of $$\overline{I}_o = \frac{1}{4\pi} \int_0^{2\pi} \int_0^{\pi} I_o(\beta,\gamma) \sin\beta\, d\beta\, d\gamma, \tag{3.16}$$

These terms describe the intensity and polarization of the occurring γ-radiation. In the case of equation (3.14), there is an anisotropy of the mean squared displacement and the intensity ratio of (3.15) different from that of (3.16). For example, the quadrupole doublet of ^{57}Fe in a polycrystalline sample will then show two lines with unequal intensities.

This effect (GKE) was detected by Gol'danskii and Karyagin in 1964. It is more expressed for greater *f*-anisotropies and γ-energies and higher temperatures than the opposite. However, it should be emphasized that the GKE is only valid for pure polycrystalline (cubic) samples—a texture may have a similar effect on the line intensities like the GKE and can easily be mixed up. The author himself only once got in touch with this GKE (Geiger et al., 1992) in the case of synthetic almandine.

During the first years after RLM's findings, merely physicists and chemists were investigating the whole periodic system for other "Mössbauer-active" elements and nuclides. As already mentioned above, around 45 elements and 1500 nuclides have been found to date. Other important elements besides ^{191}Ir (RLM's object of research) and ^{57}Fe were ^{119}Sn, ^{121}Sb, ^{197}Au, ^{151}Eu, ^{40}K, ^{61}Ni and a lot of REEs (rare earth elements).

^{119}Sn is a metastable nuclide and decays with a half-lifetime of 245 days to another state of ^{119}Sn emitting a 23.8 keV Mössbauer line. The main disadvantage of this, also very widespread, element is the comparably broad line width under normal circumstances. This renders the evaluation of spectra stemming from Sn compounds with several crystallographically non-equivalent sites complicated, since superpositions of lines can often not be resolved. However, in general, it is possible and the author also solved a few of these ^{119}Sn spectra (di Benedetto et al., 2005).

^{121}Sb is the educt of the nuclide ^{121}Sn which decays with a half-lifetime of 50 years emitting a Mössbauer line of 37.2 keV. Sb has a certain importance in metallurgy, but also in a class of minerals that can also contain Fe and Sn, but these are merely binary or ternary compounds of different composition, the so-called skutterudites. These have gained scientific attention mainly due to their thermoelectric properties. This effect refers to phenomena where either a temperature difference in the compound creates an electric potential or vice versa. Depending on the "direction" of the phenomenon we distinguish between the Seebeck effect (transforming temperature to current), Peltier effect (current to temperature) and Thomson effect (conduction by heating or cooling). In most materials these effects are too small to be technically useful. However, low-cost compounds with a comparably strong thermoelectric effect like the skutterudites might be used in applications concerning power generation or refrigeration—a commonly used material is bismuth telluride (Bi_2Te_3). The author once investigated ternary skutterudites (Grytsiv et al., 2002) by Mössbauer spectroscopy, however only the ^{119}Sn contents.

Scientifically most promising would be a Mössbauer investigation of a compound with three Mössbauer-active components like Sb, Fe, and Sn in the same sample providing a maximum of complimentary information.

^{197}Au is the educt of a platinum Pt foil activated in a nuclear reactor. The foil has a half-lifetime of only 20 h, so it is advisable to have the spectrometer close to the reactor, otherwise the available Mössbauer line of 77.3 keV has disappeared before the accumulation of an evaluable count rate. Another severe disadvantage is the low *f*-factor of the activated Pt foil—hence the Mössbauer source has to be cooled down to very low temperatures in order to enhance the intensity of the line. This can only be done in a combined cryostate, where it is possible to cool down the absorber *and* the source. This sets very high demands to the abilities of the equipment as well as of the researcher. For example, when putting the source into the He bath, ice on the foil might cause an asymmetry of the driver signal, so that the resulting Mössbauer spectra will be distorted.

However, these difficulties overcome, the Mössbauer gold spectroscopy may lead to valuable results concerning the direct neighbours in the gold lattice or even the origin of the gold contents of the sample from an archaeological point of view. Very profound studies in this respect have been performed by Wagner et al. (2000). Figure 3.10 shows a periodic system with the up to now detected Mössbauer-active elements.

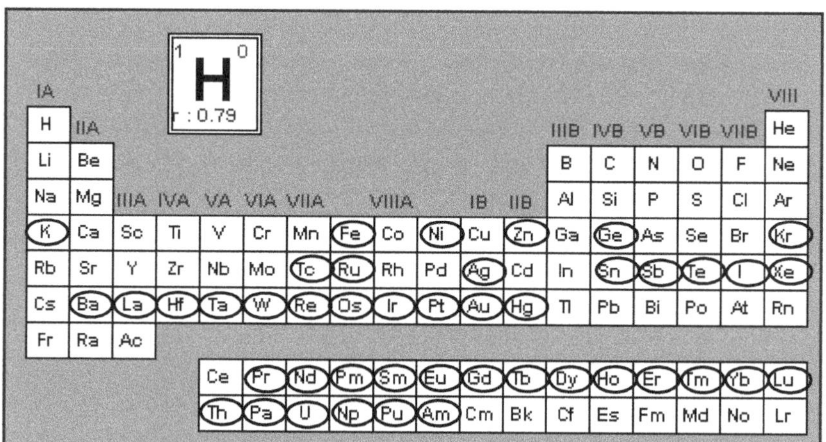

Figure 3.10 Periodic System of elements according to Mendele'ev. Mössbauer-active elements are circled.

The remaining Mössbauer-active elements belong more or less to the REEs and are only of academic interest, since their half-lifetimes are generally very low. Another aspect is economy. Even the most widespread ^{57}Co source for ^{57}Fe Mössbauer spectroscopy nowadays amounts to costs of around \$6000 (without taxes). For the production of these sources a nuclear power plant and a very well equipped radiological laboratory is compulsory. This is nowadays only undertaken by very few nations in the world, the most prominent one being Russia. Due to the very low common demand of "exotic" sources, e.g., a Mössbauer REE source would cause unrealistic high efforts and expenses. So most of these sources have been produced at the beginning of the "hype" around the effect and were not manufactured again.

In the course of time, other fields of scientific research profited from the great benefits of Mössbauer spectroscopy—among them chemistry, solid-state physics, mineralogy, material science, metallurgy, biology and medical sciences and archaeology.

A very good and detailed overview of the widespread activities is given in the monograph of Barb (1980) and, more detailed, in the book of Gonser (1975).

Here, only very few examples without the least claim to completeness shall be given in the following:

(c) Chemistry

The evaluation of the Mössbauer parameters together with dependencies on temperature, pressure, magnetic fields a.s.o. can contribute a lot to chemical properties of the given sample. The data may be explained from the electronic structure of the molecules or building units and combined with other methods like optical spectroscopy or wet chemical analysis. In particular, it is possible to attribute the isomer shifts to different oxidation states of the relevant Mössbauer atom; the most important example is the most exact determination of the Fe^{3+}/Fe^{2+} ratio as well as other states of iron with an accuracy of (relative) 1%. This is hardly achievable with wet chemical analyses. Symmetry arguments derived on the basis of the observed quadrupole splittings (e.g., QS = 0 under cubic symmetry) can help to identify the crystallographic site of the relevant nucleus and can provide important and very detailed information about the electronic configuration and the ligand distribution around the considered Mössbauer nucleus. So the grade of ionic or covalent bonds can be predicted in great detail, as well as coordination numbers (in general it is possible to distinguish—based on the concrete value of the isomer shift—between tetrahedral and octahedral surroundings.

In the case of our special example compound, synthetic fayalite (cf. Chapter 5), we were able to attribute the possible subspectra and their Mössbauer parameters to the two relevant crystallographic sites, based on the different values of the isomer shift and angle parameters—the M1 position should be more covalent due to its low site symmetry (–1) than the M2 site (*m*). Other chemical issues that can be well studied by MBS are the phase analysis (however, there exist more suitable methods like X-ray powder diffractometry XRPD), chemical reactions and catalysis. Most of these possibilities have been probed in the early years of MBS.

(d) Solid-State Physics

This scientific field certainly comprises one of the widest applications of the MBS. This is due to the fact that SSP is part of many other disciplines and it is often difficult to distinguish between those and the "pure" part of SSP. To make things clearer, it is perhaps helpful to use a widely acknowledged definition of SSP, namely: *The SSP includes all atomistically explainable phenomena of crystals and other solid states and is theoretically described by mechanical, electronic, magnetic and other excitations and their interactions.*

Because of its close connection to MBS, it is easy to understand that issues of SSP had preferably been studied in the early years of MBS. The wideness of the field, however, makes it impossible to treat all relevant topics of SSP—so we have to confine ourselves to a small, exemplary part of the early research—magnetism. One of the first issues here was to detect the magnetic hyperfine field of Fe-metal and to derive a connection to the magnetic moment of Fe. With a value for $H(0)$ of -33.0 ± 0.3 T one has detected a magnetic moment of 2.2 μ_B per iron atom (due to the polarization of 1s, 2s and 3s electrons by 3d ones). By this, it is possible to derive a proportionality factor of -150 kOe/μ_B between internal magnetic field and the mean value of the magnetic moment.

Within other samples, e.g. our main example fayalite (Chapter 5), it is—due to the fact that $H(0)$ comprises different components and due to the polarization effect by 3d electrons—difficult to predict a concrete magnetic moment and/or internal magnetic fields, respectively. In fayalite, low temperature values of $H(0)$ on M1 and M2 are around 32 and 15 T, respectively, whereas the magnetic moments are not thus different. An explanation might be that the components of $H(0)$ (H_{orb}, H_{dip} and

H_c) may in one case add up, in the other one partly compensate each other. However, these issues could not and cannot generally be explained. For this purpose, foreign atoms in Fe matrices and Fe impurities in metallic and non-metallic alloys have been investigated, where interesting results could be found concerning the coupling mechanisms between the magnetic ions, e.g. superexchange or RKKY interactions.

(e) Metallurgy

It is highly evident that ^{57}Fe Mössbauer spectroscopy is a very useful method for the examination of steels. It is possible to derive the influence of plastic deformation, mechanical and thermal processing, phase analysis and influence of other elements in steel alloys, namely carbon and silicon.

The very well-known phase diagram of steel is rather complicated—principal phases are austenite (γ-Fe) and martensite (α-Fe) with only very modest solubility of carbon in the equilibrium; martensite is metastable and can be transformed from austenite during "hardening". Austenite (γ-Fe) has an fcc structure with carbon on intermediate sites with octahedral symmetry. Carbon leads to an extension of the austenitic lattice and, hence, to a distortion of cubic symmetry. In bcc-martensite these octahedra only have one four-fold (tetragonal) symmetry—the contents of carbon linearly influence the martensitic component. In a Mössbauer spectrum on martensite at least three magnetic subspectra due to different iron-carbon environments can be distinguished, and a central line is ascribed to residual austenite. With high accuracy, much higher than chemical or mechanical methods, MBS can even contribute to the question of exsolution of carbon during aging. The influence of mechanical tension can be examined as well as thin sheets with different magnetic properties with respect to bulk material. All these examinations were performed in the late 1960s or early 1970s.

A very "useful" application of MBS is corrosion detection. It is possible to distinguish oxide phases of iron (e.g. β-FeOOH, γ-FeOOH) from the bulk material via MBS. The main advantage is that these examinations can be performed without any damage to the material, even beneath cover sheets of colour and other amorphous materials—concretely for wings of aircraft.

The author and co-workers also once examined a steel phase, namely cementite (Fe_3C) as the inner fibre of multi-walled carbon nanotubes. Beneath the broad background noise amount of the latter, a tiny, but significant hyperfine pattern could be detected in the MB spectrum (Lottermoser et al., 2006). Comparison with a spectrum of white cast iron unambiguously let the fibre material be identified as cementite. By conventional methods (e.g. HRTEM), only vague results concerning this point could be presented.

(f) Mineralogy

The main example discussed in this book, synthetic fayalite Fe_2SiO_4 (cf. Chapter 5), has been treated very early by Kündig et al. (1967). The sample belongs to the olivine group silicates $(Mg_xFe_{1-x})_2SiO_4$ with isolated tetrahedra as building elements. It crystallizes in the orthorhombic space group Pnma with 4 formula units per unit cell. The here-relevant iron cations occupy two non-equivalent crystallographic sites M1 with local symmetry −1 and M2 with *m*. This special Fe ion structure together with their surrounding oxygen octahedra gives rise to a multitude of interesting physical properties, among them a transition paramagnetic-antiferromagnetic (AF) at $T_N \approx 65$ K. There, the M2 Fe^{2+} magnetic moments order in a collinear AF pattern //**b** (Pnma), the M1 spins in a similar manner, but with an increasing canting below $T = 50$ K with decreasing temperature. At $T = 23$ K a second transition had been observed which could not be explained so far. The study of Kündig et al. examined the Mössbauer powder patterns for different temperatures below T_N and found two sets of parameters which should be attributed to both crystallographic sites. Strangely enough, the angle parameter describing the supposed canting on M1 remained independent on temperature. Our own studies on fayalite single crystals finally solved this widely discussed problem in the relevant literature: The angle parameter θ (see above, Section 3.1.1) has been found to be insensitive to the canting, whereas ϕ, which was set to constant 90° in the early work of Kündig et al. as a constraint, showed a distinct temperature dependence. In our study of 1996 (Lottermoser et al. 1996), we could even correlate previous neutron scattering results with our Mössbauer findings and solve the discrepancies that occurred. Later on, we combined all our fayalite data accumulated so far and invented the Difference Electron Nanoscope as the main topic of this book (cf. Chapter 4).

Like fayalite, an overwhelming number of iron-, tin-, antimony- and gold-containing minerals have been investigated by Mössbauer spectroscopy preferably in the last decades of the previous and the first ones of this century. Although the old "gold digger times" have ceased a bit, the boom is still going on. A very good overview of most investigated compounds is given in the Mössbauer database edited by Stevens (1998), now hosted by the Dalian Institute of Chemical Physics (DICP) at Dalian, China.

(g) Material Sciences

This topic is largely connected to the mineralogical application of the MBE, since many findings about technically applicable materials stem from natural (mineral) precursors. As an example, I mention herewith the mineral triphyllite $LiFePO_4$ which was also investigated by Mössbauer spectroscopy in 2015.

$LiFePO_4$ also crystallizes in the orthorhombic space group Pnma with four formula units per cell and is isotypical with the olivine structure. The Li is surrounded by a distorted oxygen octahedron (M1 site), Fe occupies the M2 position with less distorted octahedra and P is surrounded by O^{2-}-tetrahedra (Streltsov et al., 1993).

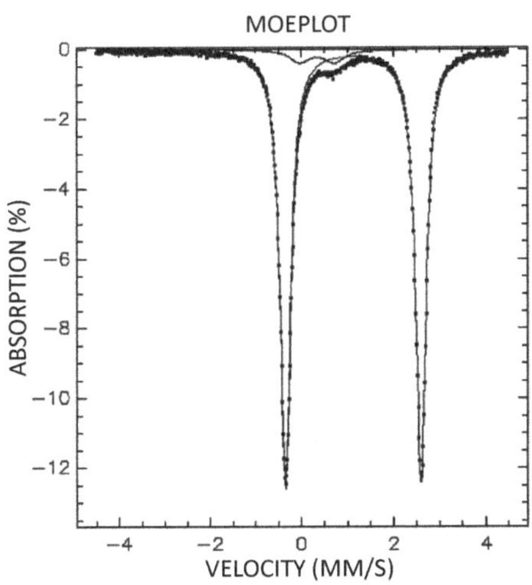

Figure 3.11 Mössbauer spectrum of triphyllite $LiFePO_4$ (dots), refined with a Fe^{2+}- and a small Fe^{3+}-doublet (straight lines).

Later on, this mineral was found to be a splendid material for batteries, preferably when Fe and Mn are combined in the lattice. The charging and discharging of the material is connected with the change of the iron valence from 2+ to 3+, which can be easily detected by Mössbauer spectroscopy. Here, a spectrum of a synthetic triphyllite for a non-industrial purpose is shown, albeit the author has investigated both, the mineral and the material (Fig. 3.11).

(h) Biology and Medical Sciences

For a long time, it has been established that biological functions of iron-containing proteids, like transport and storage of iron, transport of oxygen and electrons, catalytic functions a.s.o. can be evaluated by MBS with high accuracy. The objection might be raised that a fluid like blood, the main transport medium, could not be examined by Mössbauer spectroscopy. This is true, but—the biological active substance might be frozen in order to become a solid matter; in this state, also blood can be investigated by MBS with good results. An obstacle is the very low content of iron in the relevant molecules (which can be compensated by enrichment of the sample by pure ^{57}Fe) and the freezing procedure which must be performed very carefully in order to avoid artefacts. A nice example of the benefits by MBS on the evaluation of transport phenomena in the human body is the hem-molecule. During respiration, molecular oxygen is bound to these proteids. In its centre, it contains an iron atom surrounded by 4 nitrogens in a planar structure. Below the molecular plane, another nitrogen provides the connection to histidine thus producing the well-known haemoglobin. Above the plane, at the sixth position, the oxygen (or other light molecules like—unfortunately—CO) is weakly bound and can be easily released at a lower partial pressure than in the lungs, into the tissue thus enabling oxidation reactions there. The iron in this oxyhaemoglobin contains low-spin Fe^{2+} ($S = 0$) in the diamagnetic state (no magnetic hyperfine interaction!). These findings are derived by MBS via its low isomer shift typical for low-spin Fe^{2+}. After release of oxygen, the Fe^{2+} slightly leaves the molecular plane and gets into the high-spin state ($S = 2$), clearly visible by altered Mössbauer parameters.

The latter phenomena are the "normally" occurring blood transforming reactions. When the blood displays a certain pathology, however, the iron in hem may transform to a low-spin ferric (Fe^{3+}) ion ($S = 1/2$), again well distinguished by MBS through the Mössbauer parameters. Even complex changes of the spectrum by leukaemia and other pathologic changes can thus be established.

The above examples can almost arbitrarily be extended. However, since the present book should not be a review on MBS only, we stop at this point with the assurance of having a manifold experimental method with a wealth of possible applications at hand.

3.1.2 Single Crystal Mössbauer Spectroscopy (SCMBS)

This method is closely connected to the difference electron nanoscopy since only with the orientation of the efg within the crystallographic axes we are able to correlate other physical properties (e.g. internal magnetic fields) with structural peculiarities. The main difference between SCMBS and powder spectroscopy lies in the distribution of the line intensities. This may be best demonstrated at the example of the six line magnetic hyperfine pattern of α-iron. As shown by Grant (1975), these line intensities are strongly dependent on the direction of the incoming γ-rays with respect to the axes of the efg. By aligning this direction parallel to a lattice vector (by cutting a single crystal section with its plane perpendicularly to this lattice vector), it is possible to refine the internal magnetic field with respect to the crystallographic axes. The connection between all these directions is shown in Fig. 3.12.

Normally, it is necessary to cut at least three oriented sections of a single crystal in order to unambiguously find the axes of the efg with respect to the lattice vectors—the latter normally are principal directions in the crystal, e.g. **a**, **b**, **c** for an orthorhombic specimen.

Compared to Mössbauer powder spectroscopy, two additional angles β and α (cf. Fig. 3.12 above) can in favoured cases be refined which directly influence the intensities of the single crystal spectra.

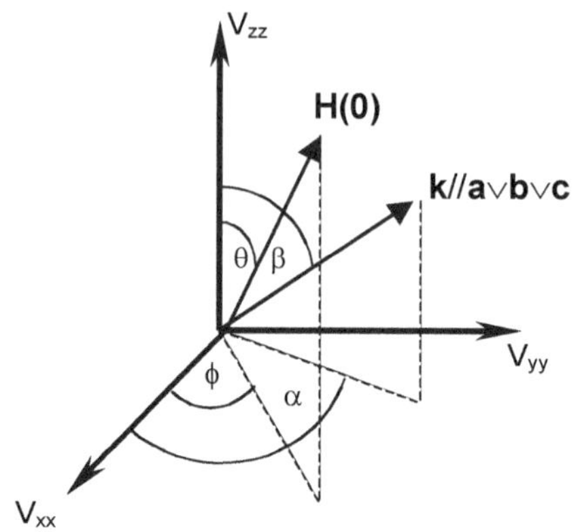

Figure 3.12 Tripod of the axes of the electric field gradient (efg) with connected k-vector of the incident γ-rays and the vector of the internal magnetic field $H(0)$ with angle convention.

However, the evaluation of these angles is not as easy as it might seem initially. It is not sufficient to prepare one crystal section with its plane perpendicular to a certain lattice vector, to record the spectrum belonging to it and evaluate the Mössbauer parameters including the orientational angles. This procedure would provide ambiguous results. In fact, at least three sections with possibly their lattice vectors strongly inclined against each other have to be provided. For example, in the case of an orthorhombic crystal, it is most useful to prepare three sections with the lattice constants and their vectors **a**, **b**, **c** parallel to the direction of the incident γ-rays. The result of the spectra refinement should then be a 90° tripod of the **a**, **b**, **c** axes with the corresponding (β, α) angles within the axes of the efg according to the notation of Fig. 3.12 above.

It is most convenient to display all this in a stereogram with the efg axes (which always form a 90° tripod) forming poles at the equatorial plane and the centre of the drawing and the axes **a**, **b**, **c** being relative to these (Fig. 3.13).

Much more useful, however, is to make the inverse drawing with the recalculated angles, so that the axes of the efg are plotted with respect to the directions of the lattice vectors **a**, **b**, **c** (cf. Fig. 3.14).

In such a stereogram with the crystal axes being the reference direction, one can also plot the poles of the internal magnetic field $H(0)$, as it has been done in our publication on synthetic fayalite (Lottermoser et al., 1996). Here, one may easily observe the temperature dependence of the efg axes and the internal magnetic field with respect to the crystallographic axes for both non-equivalent sites. The direction of $H(0)$ may be identified with the orientation of the magnetic moment from neutron diffraction measurements (cf. Section 5.6).

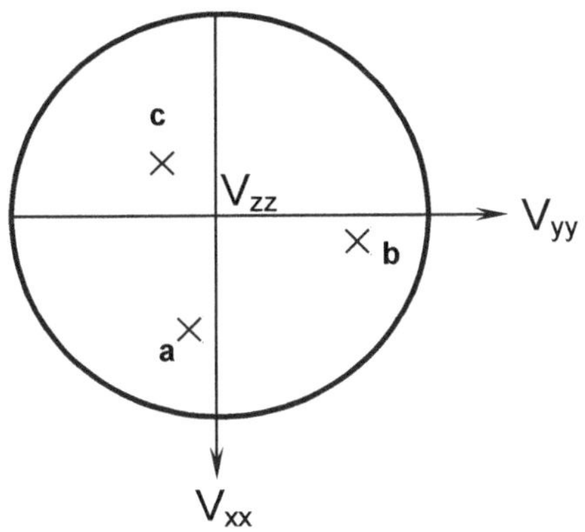

Figure 3.13 Stereographic projection of the poles of the crystallographic axes with respect to the poles of the electric field gradient efg.

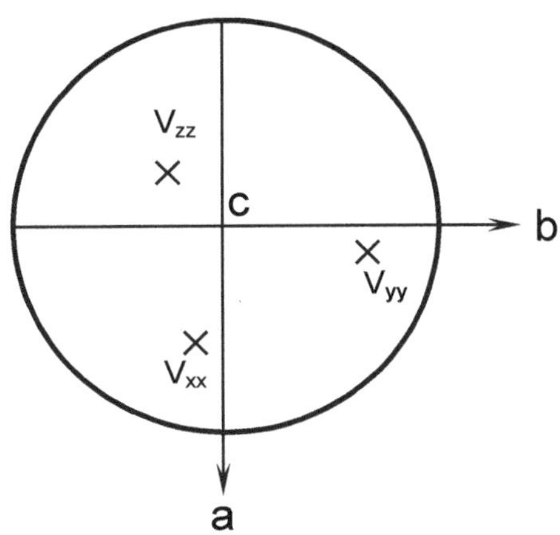

Figure 3.14 Stereographic projection of the poles of the efg axes with respect to the poles of the crystallographic axes of an orthogonal system.

3.1.2.1 Preparation of the single crystal sample

This procedure is by far much more sophisticated than the preparation of a Mössbauer powder sample; hence, it needs a good deal of peculiarities to obey before obtaining a measurable single crystal sample.

First, it is decisive, how big the "raw" specimen can be under optimum conditions. Desirable for the final crystal is a plate of about 5 × 5 × 1 mm^3, the original specimen needs to be a bit larger than this size. Smaller sizes for the final crystal are possible, but it should be taken into account that the larger the plate the stronger the intensity of the absorption peaks will be according to a rule of thumb. The hole in the standard absorber ring corresponds to this optimum value given above, smaller crystals should in any case be flat and masked by a sheet of lead in order to suppress incoherent background scattering. This mask should be fitted to the outer shape of the crystal. For example, see Fig. 3.15.

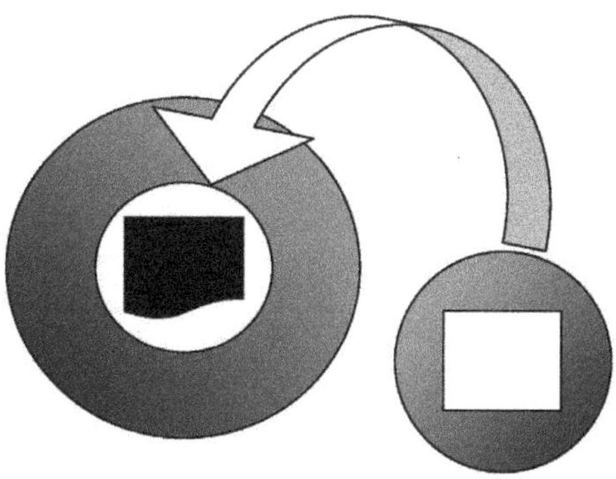

Figure 3.15 Schematic view of a Mössbauer absorber: copper ring with a flat single crystal sample (black) and Pb mask to cover the inner hole of the ring allowing mainly coherent resonant γ-ray transmission through the crystal and shading the incoherent amounts.

Second: The raw specimen should be tested whether it is really a single crystal or a twin, triple crystal or else. This may be done by recording a first X-ray Laue pattern of the raw material. Twins and other multiple crystals show in most cases a "garland" of reflections around the principal ones so that twins and else may be easily identified. If this is the case, further investigations at this specimen should be avoided since the future crystal also will have doubled or tripled efg's providing only poor and ambiguous results.

Third: Is the raw specimen transparent or opaque?

In the first case, even with a completely unshaped specimen without faces, an analysis with polarization microscopy might be eminently helpful. For this purpose, the specimen is brought onto the stage of a polarization microscope with crossed polarizers. Turning the stage will then result in darkening and lightening of the whole specimen or of regions therein. In the first case we may conclude of having a well-grown single crystal, and in the second case it will be a twin or multiple crystal with high probability, so that we may either try to cut out the region with uniform extinction (thus getting a small piece of a single crystal) or drop the raw specimen. Even "cutting out" may cause problems since there are twins which are "intrinsic" so that it is not possible to separate the two single individuals.

The first case is to be preferred in any case.

Opaque crystals only can be examined by the Laue technique or comparable methods (see below).

Fourth: Does the raw specimen show free-grown faces?

There are big volumes of mineralogic treatises on the outer shape of almost every crystal specimen (e.g., Klockmann-Ramdohr, 1942). There, it may be looked up which crystal face will be preferred at the given compound. These details give valuable hints (not always unambiguously, though) for the possible Miller indices of the selected crystal face(s). Starting from these *hkl*'s, the indexing of the future Laue patterns is considerably relieved. Even faces of synthetic crystals may be indexed since the crystal growth in special devices (e.g. after the Czochralski method) commonly shows preferred lattice directions or low-indexed crystal faces, respectively. Perpendicular to these selected directions the desired crystal plates may be cut or, by means of the arc sledges of a goniometer head, other preferred directions might be found using trigonometric relations. The desired crystal plate can then be cut perpendicularly to these new directions.

3.1.2.2 Orientation of the single crystal individuals

Let us assume that we dispose of a single crystal specimen (the "raw" sample or "rock"[6]) of sufficient size. This "rock" should be firmly attached to the top of a special goniometer head with

[6]It is called "rock", because, compared to the tiny crystals used in normal X-ray or synchrotron diffractometry, these specimen are of giant sizes.

extended arcs. Normally, the arcs of a goniometer head comprise ±10°, but for orientational purposes larger arcs are preferable. The reason is that within only ±10° a preferred special crystal direction can hardly be found without further information; the rock often has to be mounted in another direction. With larger arcs, the desired orientation can often be found within one go. Fig. 3.16 shows such a goniometer head with large arcs of ±20° (HUBER Ltd., Rimsting, Germany).

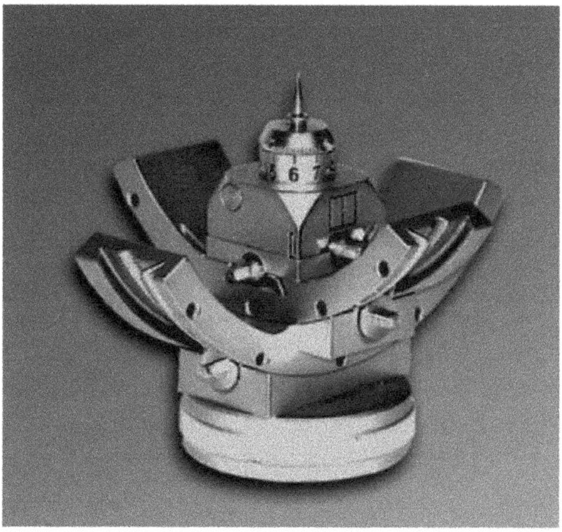

Figure 3.16 Goniometer head for the orientation of single crystals by X-ray diffraction with respect to the crystallographic axes (Huber Ltd., Rimsting, Germany). Very large arc sledges in order to allow great angle ranges.

However, it is necessary to guarantee that the rock must remain with the beam hitting its surface always at the same spot on the crystal—otherwise there may be misorientations by shifting the crystal during turning. This often may be the cause for strange observations (different crystal orientations in spite of apparently proper reverse-turning of the arcs!).

The generation of the crystal reflections by the Laue or precession methods is a section within the paragraph diffractometry (see Section 3.3.2).

Here, we may assume that we already dispose of the necessary devices like the large arc goniometer head and a suitable X-ray source with evaluation unit.

The latter is, in our case, the low-power device XR 4.0 Expert™ with Mo tube by PHYWE Ltd., Göttingen, Germany.

This apparatus is a full-protection device with a self-adjusting X-ray tube chamber so that the direction of the primary beam does not need to be calibrated. A tube can even be exchanged without further precautions since all the available tubes (Cu, Mo, W) are air-cooled. In spite of the low power of the tube (35 W at maximum), crystal orientations of almost every kind of specimen are possible.

For this purpose, a "working chamber" at the site of the primary beam with a comparably large volume and windows from lead glass are available. This compartment has a sliding door which is automatically locked during X-ray tube operation. At the left wall, the opening of the tube chamber can be covered by a cylindrical collimator in order to generate a narrow and parallel X-ray beam with a diameter of alternatively 0.5, 1 or 2 mm.

The main advantage of the device is its flexibility. At the bottom of the compartment, there is an optical bench where different tools can be installed, e.g. the goniometer head mentioned above, a fluorescent screen and a digital camera (Fig. 3.17).

Figure 3.17 Full-protection X-ray diffraction device XR 4.0, adapted to Laue exposures with molybdenum tube by PHYWE Ltd., Göttingen, Germany. On the left hand side the goniometer head with the crystal on top, the fluorescent screen and the digital camera (Canon 350D) for the recording of the Laue reflections can be seen.

Very conveniently, at the back wall of the compartment several plugs for the electrical connection and remote command of the "internal" tools are present, together with a screened opening for a non-standard wiring (cable tunnel). This open architecture allowed us to develop a recording system for Laue patterns, which is without doubt cheap (Jakhu and Lottermoser, 2016).

In former times, the reflections stemming from different net planes (cf. Section 3.3.1) have been recorded by films, later by counter tubes and recently, by image plates or electronic films. These detection methods have in common that they are comparably expensive or uncomfortable, respectively, since the use of chemicals will be necessary in the conventional film case. The erstwhile widely spread and comparably cheap polaroid™ films have almost disappeared from the market. Since it was impossible for us to afford an image plate detector, we developed the above-mentioned system:

The primary beam hits the sample crystal (control by an inversely mounted LASER beam and a movie cam) and produces a Laue pattern on a fluorescent screen. The reflections are transformed into visible light and recorded from the backside with a digital camera and long-term exposure. Some examples of these Laue patterns of different crystals are shown in Fig. 3.18.

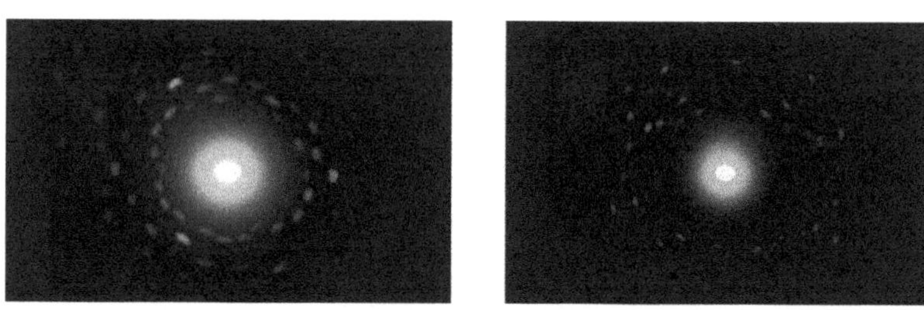

Figure 3.18 Different Laue exposures with the principal symmetry axis in the centre of the screen. Left: Laue pattern of quartz with the X-ray beam parallel to the crystallographic *c*-axis, trigonal symmetry clearly visible. Right: Fayalite Laue pattern with the *b*-axis parallel to the X-ray beam, rhombic symmetry, only slight misalignment.

We started with the test crystal LiF attached to the basic equipment of the X-ray device and were quickly successful with our low-cost recording. In particular, in the case of quartz, it was easy to find the main crystallographic axis of the "rock" by visual inspection and subsequent cut and found a nice triple symmetry on the Laue pattern (Fig. 3.18, left).

Our example crystal synthetic fayalite was previously oriented on another device (Buerger precession camera, see Section 3.3.2.1), but this test crystal, a plate of ideal size, nicely shows the expected Laue pattern (Fig. 3.18, right).

However, in cases where the orientation of the "rock" is not clear, the desired main axes have to be detected by inspection of the Laue pattern and a crystallographic visualization program (e.g. CaRINe by Boudias and Monceau (1989–1998)). To find the orientation may be very arduous in these cases. It is recommended to use an online Laue program like LAUEGEN by Weber (1997). Here, several basic parameters like lattice constants and angles, crystal system, space group and range of desired *hkl* can be put in. From these details, LAUEGEN creates a theoretical Laue pattern with the direction of the suggested lattice vector parallel to the primary beam. By changing the *hkl* of this supposed lattice vector it is possible to subsequently produce new Laue patterns till the theoretical image fits well to the observed recorded pattern.

By help of the crystallographic visualization programme (CaRINe™ is capable of performing calculations of angles between lattice vectors), it is possible to detect how the "rock" should be turned in order to get to the desired main direction.

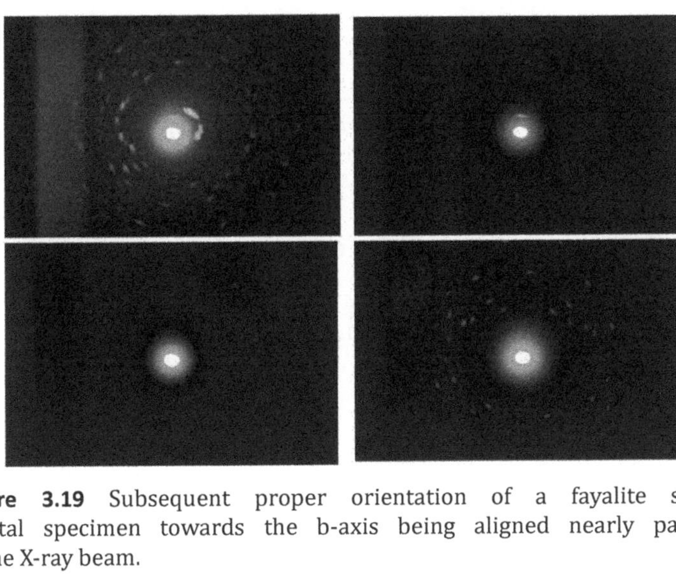

Figure 3.19 Subsequent proper orientation of a fayalite single crystal specimen towards the b-axis being aligned nearly parallel to the X-ray beam.

The latter can be very often found at once by recognizing the commonly enhanced symmetry (mirror plane or else) after adjustment of the relevant arcs at the goniometer head.

The procedure is shown in Fig. 3.19, where several orientations of the fayalite crystal close to the primary beam //**b** are shown.

In Fig. 3.20 we also see the corresponding theoretical Laue patterns created by LAUEGEN (Jakhu and Lottermoser, 2016).

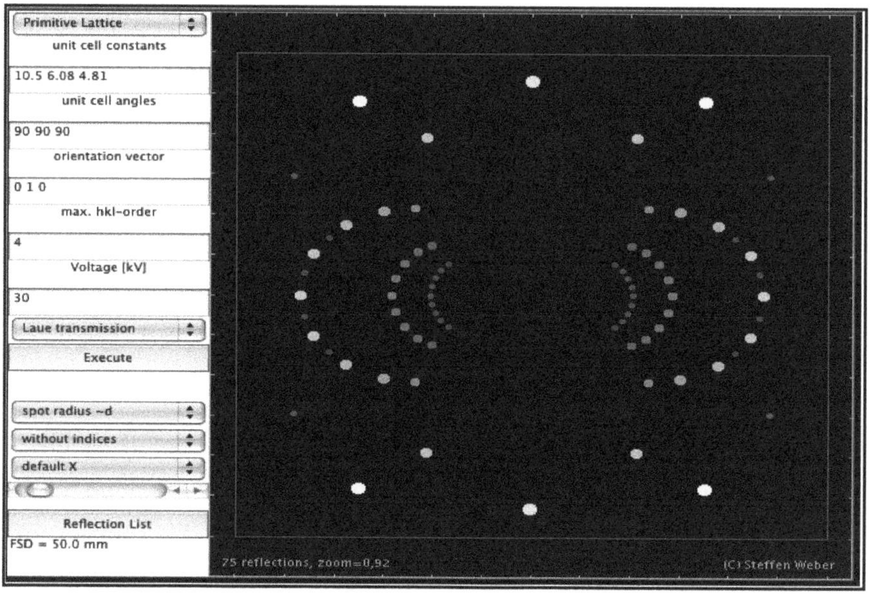

Figure 3.20 The crystal orientation program LAUEGEN by Steffen Weber with the simulated Laue pattern containing the X-ray beam parallel to the b-axis. Good correspondence with the last exposure (experimental) of Fig. 3.19.

3.1.2.3 Manufacturing of the oriented single crystal samples

Having found the desired lattice direction by applying the above procedure, the goniometer head together with the oriented rock is mounted on a special stage; the rock is attached to a glass plate by using an epoxy glue and—after drying—carefully removed from the wax drop of the goniometer head. The edges of the glass plate should be aligned parallel to the intended cut directions of the crystal saw.

For the cutting device, several commercial and in-house built equipments are commonly used. Most widespread is a wire saw, where a fathom covered with grinding material performs the cut—the rock is mounted on a lever with an adjustable counter mass, so that the pressure of the fathom on the rock can be changed according to the hardness of the material. By this device, it might be stated that the crystal is cut by its own weight. Additionally, it is possible to lead the fathom through a water bath so that optimal cooling of the saw is warranted.

We, however, developed our own cutting device since there is almost no feedback from the saw to the user during cutting with conventional devices. Our basic machine is a mini-milling table with X–Y increments of 0.05 mm at minimum (cf. Section 5.2). In the Z-direction a circular diamond saw on a comparably strong mini-motor adjustable in rpm can be manually lowered by a long lever. So it is possible to either cut the rock very sensitively by help of the lever or to fix the position of the saw and adjust very carefully the manual proceeding of the arcs of the X–Y table. Additionally, the circular saw can be cooled by a fixed sponge with a special fluid. With the above device, we always have a very sensitive feedback from the crystal whether the rpms are too high or too low, the increment of the cutting is suitable or the pressure to the crystal appropriate. Hence, we mostly have gotten very good crystal sections, be it from soft, brittle rocks or be it from very hard material up to Mohs hardness 8.5.

The cut itself is performed by fixing the glass plate with the oriented rock on the milling table aligned with its edges or special marks parallel to the plane of the circular saw, putting the blade close to the plane of the glass plate and to cautiously move the X-arc of the table. Having performed a cut in this X-direction, the blade is raised and the Y-increment put forth or backwards by an amount corresponding to the desired thickness of the crystal section.

This thickness depends on several parameters like hardness of the crystal, future measurement a.s.o. and should be chosen carefully. A minimum thickness of commonly 1 mm is recommended for most purposes, since it is easier to grind the crystal plate to a lower thickness than to replace a broken crystal—this may happen easily if the cuts are too close to each other.

When the circular saw touches both times the glass plate below the crystal plate, one may carefully try to remove it. A sharp knife can be used to scrape the last bit of glue or wax from the glass plate in order to free the new crystal plate.

Having removed the latter, it can be fixed by a heated and dried drop of wax (hot plate!) with its plane on a second glass plate—if necessary, edges of the crystal plate may be cut parallel to each other and the residual slivers removed.

For purposes of neutron diffraction measurements, the crystal plate may be ready for the experiment and glued with one of its edges on a special sample holder. The orientation of the crystal section may again be verified on the X-ray device.

For purposes of Mössbauer spectroscopic measurements, the crystal plate has to be very thin in order to get recordable resonance scattering. Commonly, the plate might be grinded to a thickness of about 0.1 mm ("thin section").

For this purpose, we dispose of a modified turntable where the lever arm holds the glass plate with the crystal by help of a suction bowl. Emery paper of different abrasive grades can be fixed on the turntable. By rotation of the turntable with the lever arm with the crystal lying on the emery paper, the crystal plate can be ground or polished to nearly every thickness. In the middle of this working procedure, the crystal plate should be removed, put into an absorber ring (as described above) and measured on the Mössbauer device (reduced count rate) in order to verify whether the crystal is transparent for γ-rays. If not, the plate has to be ground again to a new thickness. This arduous procedure may comprise several cycles. When the crystal plate gets "open" (transparent for X-rays), it may be fixed to the aluminium backsheet of the absorber ring and measured with full count rate.

With at least three of these thin sections with their plane perpendicular to different lattice vectors (e.g., **a**, **b**, **c** in the orthorhombic case) the refinement with the additional orientational angles β and α has to be performed and the angles evaluated. As mentioned above, the angles must fit in such a manner to each other that the axes of the resulting efg will form a 90°-tripod. This constraint is a challenge for the Mössbauer researcher since the angles of one section are often ambiguous and it is arduous to find the correct ones for all three thin sections together. A small supporting routine for calculating the possible angle combination which leads to the correct efg tripod may help a lot in this situation.

In the case of synthetic fayalite, we have found one unambiguous angle combination for each crystallographic site (see Table 5.2, Lottermoser et al., 1996).

3.1.3 Nuclear Magnetic Resonance (NMR) and Nuclear Quadrupole Resonance (NQR)

As already mentioned above, the main disadvantage of Mössbauer spectroscopy lies in the fact that it is confined to special elements or nuclides, respectively. However, there is a possibility to produce the Mössbauer effect by synchrotron radiation avoiding the above restriction, but this method has never gained the great importance of the "original" Mössbauer spectroscopy.

So the question remains whether there are other methods to especially provide an efg independent of the few nuclides that are Mössbauer-active.

Indeed, there are: They also use nuclear resonance effects, but not due to γ-rays but due to transition between nuclear energy levels which are split by an external magnetic field or just without the latter.

3.1.3.1 Basics

Let us consider a nucleus with a magnetic moment μ_N and an angular momentum $\hbar\mathbf{I}$. Since these quantities are parallel, we may write

$$\vec{\mu}_N = \gamma \hbar \vec{I}, \tag{3.17}$$

where γ is the gyromagnetic ratio and $\mathbf{I}$ the nuclear angular momentum.

The interaction energy with an external magnetic field amounts to

$$E = -\vec{\mu}_N \vec{B}_{ext}. \tag{3.18}$$

With $\mathbf{B}_{ext}$ in the z-direction, $\mathbf{B}_{ext} = B_o \, \mathbf{e}_z$ we get

$$E = -\mu_z B_o = -\gamma \hbar B_o I_z. \tag{3.19}$$

The allowed values of I_z run from $m_I = I, I-1, \ldots -I$, so we may express the nuclear spin with the magnetic quantum number.

In a magnetic field, a nucleus with $I = 1/2$ possesses two magnetic sublevels $m_I = \pm 1/2$. When the difference between the two levels is denoted by $\hbar\omega_o$, we may set $\hbar\omega_o = \gamma\hbar B_o$ or

$$\omega_o = \gamma B_o. \tag{3.20}$$

This is the condition for resonance absorption under Zeeman splitting of nuclei in a static magnetic field (Fig. 3.21).

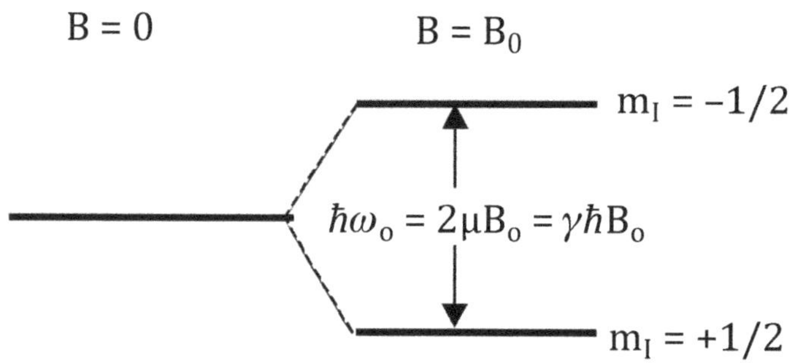

Figure 3.21 Zeeman splitting of nuclear levels in a static magnetic field B_0.

Thus, the resonance frequency of, e.g., a proton can be calculated from B_o and γ, the latter from $\gamma = 2\mu_P/\hbar$ and the value for the nuclear magneton μ_P:

$$\nu[\text{MHz}] = 42.58 B_o[T] \tag{3.21}$$

A table for the nuclear magnetons of many nuclides is given, e.g., in Kittel (1980).

Analogous to Newton's third law, the time derivation of the angular momentum is equal to the torque acting on a magnetic moment μ by a magnetic field:

$$\hbar\frac{dI}{dt} = \vec{\mu}\times\vec{B}_{ext} \quad \text{or} \quad \frac{d\vec{\mu}}{dt} = \gamma\vec{\mu}\times\vec{B}_{ext} \tag{3.22}$$

If we define the nuclear magnetization as a sum over all single nuclear magnetic moments, we obtain

$$\frac{d\vec{M}}{dt} = \gamma \vec{M} \times \vec{B}_{\mathrm{ext}}. \tag{3.23}$$

Let us assume a static external field $\mathbf{B}_{\mathrm{ext}} = B_0\mathbf{e}_Z$ in the vertical direction. We will then obtain a magnetization $//z$ in thermal equilibrium without components in M_x and M_y, whereas in the case $I = 1/2$ the resulting magnetization will be determined by the occupation number of the two magnetic sublevels (Boltzmann factors).

The magnetization in equilibrium will then be given by

$$M = N\mu \tanh(\mu B/k_{\mathrm{B}}T) \tag{3.24}$$

with $M_0 = N\mu$.

This equation is rather easy to verify. The present magnetization is dependent on the ratio of magnetic and thermal energy: A huge μB has the same effect on magnetization as a low $k_{\mathrm{B}}T$. The maximum possible magnetization will take place, when all nuclear magnetic moments are aligned: $M_Z = M_0 = N\mu$. In this case, the magnetization is saturated and in the equilibrium.

If M_Z is not in equilibrium, it will approach M_0 with a rate proportional to the deviation of M_0. In the formula:

$$\frac{dM_z}{dt} = \frac{M_0 - M_z}{T_1} \tag{3.25}$$

T_1 is the longitudinal relaxation time or the spin-lattice relaxation time. In the beginning of the onset of the magnetic field, magnetic moments have to be aligned. The result may be derived by integration of the previous equation; dependent on t/T_1 we will get a saturation of the magnetization (Fig. 3.22):

$$\ln\frac{M_0}{M_0 - M_z} = \frac{t}{T_1}; \quad M_z(t) = M_0(1 - e^{-t/T_1}) \tag{3.26}$$

The relaxation can be obtained within three possible tracks under phonon contribution.

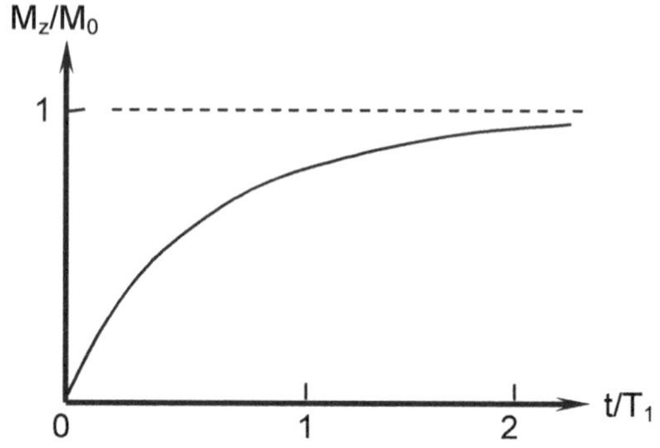

Figure 3.22 Normalized magnetization in z-direction dependent on time in units of relaxation time T_1 during a NMR experiment. Clearly visible is the time-dependent saturation of the magnetization.

This relaxation time T_1 is a main quantity within NMR/NQR spectroscopy.

Together with the relaxation amount, we get for the vertical (longitudinal) component of the motion equation:

$$\frac{dM_z}{dt} = \gamma(\vec{M} \times \vec{B}_{\text{ext}})_z + \frac{M_o - M_z}{T_1} \tag{3.27}$$

where **M** will, together with its precession motion around the magnetic field, approach the equilibrium value M_0. However, if there are transversal components M_x and M_y, they will tend towards zero and may be described by the following two equations:

$$\frac{dM_x}{dt} = \gamma(\vec{M} \times \vec{B}_{\text{ext}})_x - M_x/T_2$$
$$\frac{dM_y}{dt} = \gamma(\vec{M} \times \vec{B}_{\text{ext}})_y - M_y/T_2, \tag{3.28}$$

where T_2 is the transversal relaxation time.

This quantity is a measure for the time, how long the moments contributing to the total magnetization will remain in phase. On the different moments act different local magnetic fields which cause different precession frequencies. With all spins initially being in phase, we obtain in the course of time a statistic phase connection between them; the transversal components will disappear. T_2 may be considered as the period, within the phase relations will vanish.

The previous equations are known as Bloch relations. In the experiment, a sample will be mainly exposed to a static magnetic field $//z$ together with a high-frequency field in x or y (Fig. 3.23).

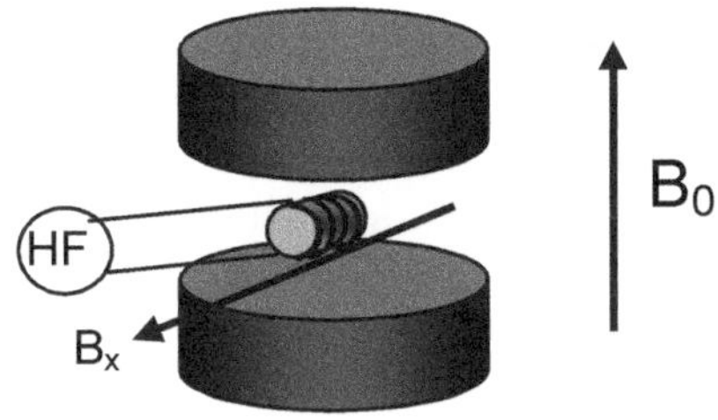

Figure 3.23 NMR experiment on a cylindrical sample with a vertical static magnetic field B_0 and high-frequency fields (HF) in x or y direction, respectively. The emission and detection coil, as well as the registration unit is symbolized by "HF".

With $\mathbf{B}_{\text{ext}} = B_o\mathbf{e}_z$ and magnetization in z-direction being constant, we obtain

$$\frac{dM_x}{dt} = \gamma B_o M_y - \frac{M_x}{T_2}; \quad \frac{dM_y}{dt} = \gamma B_o M_x - \frac{M_y}{T_2}; \quad \frac{dM_z}{dt} = 0. \tag{3.29}$$

As a solution, we assume damped harmonic oscillators:

$$M_x = me^{-t/T'} \cos \omega t \quad M_y = me^{-t/T'} \sin \omega t \tag{3.30}$$

Applied to the Bloch equations, we obtain a two-dimensional damped oscillator system (spin system) with resonance absorption of the HF-energy close to the frequency $\omega_o = \gamma B_o$ with a half width at half maximum of

$$\Delta\omega \approx 1/T_2 \tag{3.31}$$

Modern measurement methods do not use continuous oscillation fields but radio-frequency pulses. A short RF-pulse excites a frequency band whose width is by Fourier relation inversely proportional to the length of the pulse. In the sample all transitions, which belong to this band, will be equally excited. Using correct length and power of the pulses, the magnetization of the excited nuclear spins may be shifted to the plane perpendicular to the main magnetic field. After the pulse, the transversal magnetization will oscillate further on in the same direction with each nuclear spin oscillating with its own Larmor frequency. The sum of these oscillations will be recorded with the same coil as the emitted pulse and subsequently digitalized. By help of a Fast Fourier Transform (FFT) it is possible to extract the individual Larmor frequencies. The modern procedures are therefore called PFT-NMR (Pulsed Fourier Transform NMR Spectroscopy.

A disadvantage of the NMR method is its comparably low sensitivity. However, there are different tricks and methods in order to overcome this shortcoming. The main application fields of NMR are the damage-free determination of the contents of mixtures, the detection of molecular structures from small molecules up to proteins and the examination of interactions between molecules.

For our purposes, however, a special branch of the NMR can be used for nuclei with $I > 1/2$. As mentioned above, these nuclei dispose of a quadrupole moment which may interact with the electric field gradient(*s*) of the given sample. The different physical quantities playing an important role here have also been given above. In particular, by nuclear quadrupole resonance (NQR) the Hamiltonian of the interaction between the efg and the nuclear quadrupole tensor may be determined according to the following equation:

$$\hat{H}_Q = \frac{e^2qQ}{4I(2I-1)}[(3I_z^2 - I^2) + \eta(I_x^2 - I_y^2)] \tag{3.32}$$

Similar to MBS, angle parameters of the efg can be detected; of special importance is the so-called quadrupole frequency:

$$\omega_Q = \frac{\pi 3e^2qQ}{2I(2I-1)h} \tag{3.33}$$

Applying a strong magnetic field to a nucleus with a quadrupole moment, the Zeeman splitting will dominate—the quadrupole interaction is a perturbation of this splitting. The sites of the energy levels can then be calculated by perturbation theory, whereas the first order p.t. will give the splitting of the central line and the second order will lead to a shift of the central line and its satellites within the NMR/NQR spectra (Fig. 3.24).

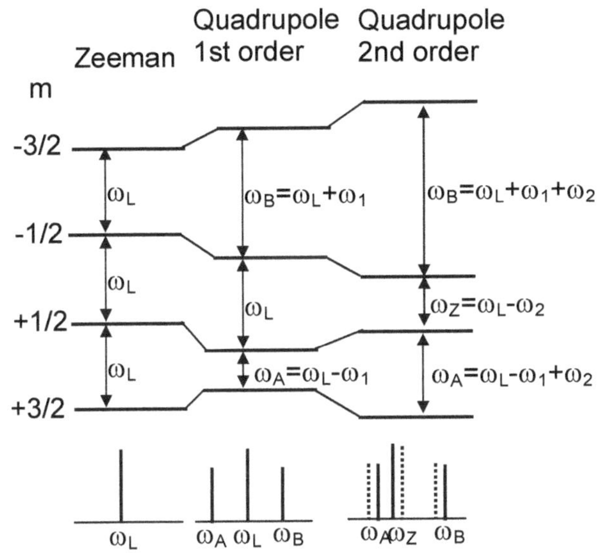

Figure 3.24 Energy levels for a nucleus with $I = 3/2$ and combined Zeeman and quadrupole interaction. m denotes the magnetic quantum number and ω the different combinations of transition frequencies. The lower image displays the correlated distribution of the different frequencies on the ω-axis with occurring shifts of the line positions.

Even the asymmetry parameter η can be determined by inspection of the outer shape of the lines—hence, also by NMR/NQR the calculation of an efg is similarly possible as in the case of Mössbauer spectroscopy, with the nice perspective that we can get the efg for some further elements than accessible by MBS.

3.2 The Full Quantitative (Supporting) Method to Calculate an efg from First Principles

3.2.1 Fundamentals of Theoretical Approaches: Density Functional Theory (DFT)

(a) Introduction

Basis of all quantum-mechanical calculations is the famous Schrödinger equation (SE), here in the time-independent and non-relativistic mode:

$$\hat{H}\varphi = E\varphi, \tag{3.34}$$

where $\hat{H}$ is the Hamilton operator comprising a term for the kinetic energy of a particle, $-\hbar^2/2m$, a potential $V(r)$ term and the eigenvalues E of this Hamiltonian; φ is a suitable wavefunction (WF) describing the probability in space of the quantum mechanical particle. Just to illustrate the effect of the SE, we will consider the easiest possible case, the motion of one free electron (without potential) using a harmonic wavefunction e^{-ikx} in one dimension. The solution to the relevant SE by input of the above WF will then be

$$\hat{H}\varphi = E\varphi \Rightarrow -\frac{\hbar^2}{2m_e}\frac{\partial^2}{\partial x^2}e^{-ikx} = Ee^{-ikx} \Rightarrow \frac{-\hbar^2(ik^2)}{2m_e}e^{-ikx} = Ee^{-ikx}$$

$$E = \frac{\hbar^2 k^2}{2m_e} \tag{3.35}$$

This is the well-known parabolic dependence of the energy on the wavevector **k** in a very simple one-electron model of a conductor in the Fermi-Sommerfeld-gas approximation.

Naturally, the solution of a SE in a molecular cluster or solid is much more complicated. The Hamiltonian with the above restrictions would then be:

$$\hat{H} = \hat{T}_{\mathrm{N}} + \hat{T}_e + \hat{U}_{\mathrm{NN}} + \hat{U}_{\mathrm{ee}} + \hat{U}_{\mathrm{Ne}}, \tag{3.36}$$

with T_{N} und T_{e} being the operators of the kinetic energy of a similar significance as above, but valid for N nuclei and K electrons and U_{NN}, U_{ee} describing nuclear-nuclear and electron–electron repulsion and U_{Ne} attractive nuclei-electron interaction, the latter three terms being of Coulomb type.

With some restrictions, the corresponding SE can be transformed into a less complicated form, namely by separation of the electronic and nuclear motion (Born–Oppenheimer approximation). In this case, the nuclear and the electronic part within the wavefunction (WF) can be separated and written as a simple product:

$$\Psi(\vec{R},\vec{r}) = \psi_{\mathrm{nuc}}(\vec{R})\psi_{\mathrm{el}}(\vec{r}) \tag{3.37}$$

with **R** being the nuclear coordinates and **r** the electronic ones.

The Hamiltonian in Born–Oppenheimer (BO) approximation can then be written as

$$\hat{H}_{\text{el}} = \hat{T}_{\text{e}} + \hat{U}_{\text{Ne}} + \hat{U}_{\text{ee}} \tag{3.38}$$

Here, the electrons move in the electrostatic field of fixed nuclei and the nuclei move in an energy surface given by the solution of the SE for electrons. U_{ee} is again the repulsive pure electronic term. The effect of this BO approximation is that a cloud of negative electronic charge forms the "glue" for the nuclei repulsion and defines surface and shape of the molecule. The total energy of the system comprises the sum of electronic energy E_{el} and the nuclear repulsion term E_N based on U_{NN}.

The BO-approach provides very good results for the hydrogen molecule. However, it does not take the Pauli exclusion principle and other issues into account. However, without the above disadvantages, it is enough for calculating a molecular energy to solve the electronic SE and to add this energy to the internuclear repulsion.

(b) Properties of DFT

The DFT is nowadays state of the art for quantum mechanical approaches to understand the physical properties of molecules and even solids.

This was not possible up to the 1970s (see below, Section 3.2.1.1). The knowledge of the electronic energies and their wavefunctions enables to calculate the magnetic, optical and thermal properties of the system under consideration and leads to a complete understanding of atomic, molecular and solid-state physics. However, the path to arrive at this goal has been long and stony. The WF is a complicated quantity depending on $3N$ spatial and N spin variables for an N-electron system. To approach this complex problem, it is necessary to introduce a single-particle density $\rho(r)$ and a two-particle one, $\pi(x_1, x_2)$. The first one is a probability density, which leads integrated to the total number of electrons N—$\rho(r)$ is an observable, it can be even determined by X-ray diffractometry (see Section 3.3.1). The last one is very important for the evaluation of electronic correlation (i.e. interelectronic interactions and influence). This term can be replaced by the product of individual probabilities (since electrons may be considered as identical particles). Hohenberg and Kohn (1964) then found that an exact description of the electronic structure can be given on the basis of the one-pair density. The general concept is older (Tomas 1927, Fermi 1927), but it was extended to the electronic systems mentioned above; it is the idea that a probability density can replace the relevant WF. In order to explain this concept of the DFT, we should cite the three Hohenberg-Kohn (HK) theorems:

(1) The external potential $U_{ext}(r)$ of the N-electron system is uniquely determined via the one-electron density (besides an unimportant constant).

This means that with the known electron density of the ground state $\rho_0(r)$ it is possible to calculate any ground state property (e.g. magnetic order at $T = 0$ K).
Next, a universal functional for the kinetic energy of electrons and their interaction is defined as

$$F[\vec{\rho}(\vec{r})] = \int \Psi^*(\hat{T} + \hat{U}_{ee})\Psi d\vec{r}. \qquad (3.39)$$

It is valid for any number of particles and any external potential.
By help of $F[\rho(\mathbf{r})]$, the energy functional for a given external potential $U_{ext}(\mathbf{r})$ is defined as

$$E[\vec{\rho}(\vec{r})] = \int U_{ext}(\vec{r})\rho(\vec{r})d\vec{r} + F[\vec{\rho}(\vec{r})]. \qquad (3.40)$$

Using WFs ψ_t and integrating the Hamiltonian H_t to give the energy E_t, it is possible—by variation of $E_t(\psi_t)$ to search all reasonable WFs ψ_t in order to find the minimum ground state energy and WF which is always lower than these "test" quantities (variational principle).
From this we may follow the second Hohenberg-Kohn theorem:

(2) The energy of an N-electron system calculated from a trial one-electron density $\rho'(\mathbf{r})$ with the properties $\rho'(\mathbf{r}) > 0$ and $\int\rho'(\mathbf{r})d\mathbf{r} = N$ is always larger than the exact ground state energy E_0:

$$E[\rho'(\mathbf{r})] \geq E_0[\rho_0(\mathbf{r})] \qquad (3.41)$$

Again: The external potential (describing properties of the many-particle system) is determined by the ground-state electron density distribution.

The solution of the problem can be obtained by application of the third HK-theorem:

(3) The total energy of an N-electron system in a local, external potential $U_{\text{ext}}(\mathbf{r})$ can be described by

$$E[\vec{\rho}(\vec{r})] = \int \rho(\vec{r}) U_{\text{ext}}(\vec{r}) d\vec{r} + T[\vec{\rho}(\vec{r})] + U_{\text{ee}}[\vec{\rho}(\vec{r})]$$
$$= U_{\text{ext}}[\vec{\rho}(\vec{r})] + F_{\text{HK}}[\vec{\rho}(\vec{r})]. \quad (3.42)$$

F_{HK} is of the same form in all systems, so that the many-electron system can be considered as a general problem, independent on the concrete system.

However, to find the universal functional, Kohn and Sham (1965) found a suitable expression:

$$F[\vec{\rho}(\vec{r})] = T_{\text{s}}[\vec{\rho}(\vec{r})] + J[\vec{\rho}(\vec{r})] + E_{\text{xc}}[\vec{\rho}(\vec{r})]. \quad (3.43)$$

where T_{s} is the kinetic energy of non-interacting electrons, J the classical Coulomb energy from the interaction of two charge densities and E_{XC} the exchange correlation energy which may be split into two terms, the exchange E_X and the correlation E_{c} contribution, where

$$E_X \approx 10^* E_{\text{c}}. \quad (3.44)$$

All summarized, we obtain the total Hamiltonian of this complex electronic system equivalent to one-electron Kohn–Sham equations:

$$\left\{ -\frac{\hbar^2}{2m_e}\nabla^2 + \hat{V}_{\text{ext}}[\vec{\rho}(r_i)] + \int \frac{e^2 \rho(\vec{r}_j) d^3\vec{r}_j}{\left|\vec{r}_i - \vec{r}_j\right|} + \hat{V}_{\text{xc}}(\vec{r}_i) \right\} \psi_i(\vec{r}_i) = \varepsilon_i \psi_i(\vec{r}_i) \quad (3.45)$$

The WF's are molecular orbitals $\psi_i(r_i)$ and called Kohn–Sham orbitals, and V_{xc} is the exchange-correlation functional obtained as a variation of E_{XC}:

$$V_{xc}(\vec{r}_i) = \frac{\delta E_{xc}[\rho(\vec{r})]}{\delta \rho(\vec{r})} \quad (3.46)$$

The one-electron density is defined as

$$\rho(\vec{r}_i) = \sum_k n_k \, |\psi_k(\vec{r}_i)|^2, \quad (3.47)$$

where n_k is the occupation number of molecular orbital (MO) at $k, \psi_k(\mathbf{r})$.

So this density is just a sum of the probabilities of MO's being built up by these. With the knowledge of the ground-state density function $\rho_0(\mathbf{r})$ and the exchange correlation energy $E_{xc}[\rho_0(\mathbf{r})]$ we will obtain the exact energy of the system.

If we want to solve the KS equations, we may use an approach called local density approximation, which is an expansion for E_{xc} depending on a local density ε_{xc}, where the exchange part is already known by the formula of Dirac:

$$\varepsilon_x^{LDA}[\rho(\vec{r})] \propto \rho^{1/3}(\vec{r}) \quad (3.48)$$

This is valid for an electron in a uniform electron gas. For the correlation term $\varepsilon_c^{LDA}[\rho(\vec{r})]$, there is no simple functional form but it can be taken from, e.g., quantum Monte Carlo calculations.

Most of the molecules dispose of an even electron number, since they are paired (closed shell systems, described by LDA). Each spatial KS orbital is occupied by two electrons with up-spin α and down-spin β. This is a typical low-spin configuration.

Some levels, however, are occupied by one or more unpaired electrons (high-spin configuration). The spatial KS WF's ψ_α and ψ_β and their orbitals are then occupied singly. Hence, they have—contrary to the previous case—different orbital energies and different ρ_α and ρ_β (open-shell systems). This "unrestricted" LDA approach, valid for (para-)magnetic ions is called local spin density approximation (LSDA). The total probability density is then the sum of single densities ρ_α and ρ_β and the exchange energy described as the sum of the cubic roots from these spin densities individually.

The LSDA method predicts structural and ground-state properties for solids rather well. Moreover, the DFT calculations are easier to perform than the HF (Hartree–Fock) ones. The basis set requirements for DFT methods are comparable to those for HF procedures and can be considered as being ab initio.

In order to account for non-local contributions to the exchange term of the HF approach, Slater proposed to replace this non-local contribution by an approximate expression:

$$U_{\mathrm{HF}}^{x}(\vec{r}) = c_x \left(\frac{\rho(\vec{r})}{\pi} \right)^{1/3} \tag{3.49}$$

with c_X being a constant which is adjusted by a semi-empirical factor α:

$$c_x = -3\alpha \frac{e}{2} \tag{3.50}$$

Hence, this Hartree–Fock–Slater equation becomes:

$$\left\{ -\frac{\hbar^2}{2m_e}\nabla^2 - \sum_I \frac{Z_I e^2}{\left|\vec{R}_I - \vec{r}_i\right|} + \int \frac{e^2 \rho(\vec{r}_j) d^3\vec{r}_j}{\left|\vec{r}_i - \vec{r}_j\right|} - \frac{3\alpha e^2}{2}\left(\frac{3\rho}{\pi}\right)^{1/3} \right\} \psi_i(\vec{r}_i) = \varepsilon_i \psi_i(\vec{r}_i) \tag{3.51}$$

The first term is the kinetic electron energy, the second one the attractive potential between nuclei and electrons U_{Ne}, the third the repulsive potential of the electron cloud and the fourth the exchange correlation potential, parametrized by α (here: α = 0.7). The expression for the exchange energy only, Xα, is a very effective tool to reduce calculation time as compared to real ab initio HF model.

3.2.1.1 Historical background (conventional quantum mechanics)

By the beginning of the 20th century, the Bohr treatment of electronic orbits in atoms, the positive nuclear charge, atomic number and position of atoms in the periodic table was state of the chemical knowledge. However, several findings could not be understood in the frame of Bohr's theory, e.g. that an electron on a certain orbit should (according to Maxwell's electrodynamics) loose—as a moving charge—energy continuously and fall into the nucleus. These and other discrepancies could be solved in the 1920s by the development of quantum mechanics and Schrödinger's famous wave equation (non-relativistic) for matter waves:

$$\hat{H}\Psi = E\Psi \tag{3.52}$$

using Heisenberg's matrix mechanics. A relativistic extension was given by Dirac a short time later.

However, a more complicated solution than the hydrogen atom could not be provided so far; it seemed to be impossible to give polyatomic solutions—even the hydrogen molecule was only unsatisfactorily solved by Heitler and London in 1929. With the hopelessness of exact solutions, approximate approaches were developed which have, according to John A. Pople, to meet six requirements:

(1) Target accuracy—the data should be reproducible and should be predicted within experimental errors.
(2) The approximate mathematical procedure should be widely general and continuous, so that a full theoretical model chemistry could be provided.
(3) The developed method should be limited in cost and time. This aims at the effectiveness of computer programs.
(4) The model has to be verified at high-quality experimental data with exemplary significance for other compounds.
(5) The model should be able to predict phenomena whose explanation is unknown or in dispute.
(6) Whenever possible, the model should work ab initio, i.e. only with fundamental physical constants and not empirical parameters.

The approximate models developed in the 1930s are more or less of qualitative nature and let the electrons occupy independent molecular orbitals (MO theory). The basics of these many-electron wavefunctions were given by Hartree, Fock and Slater. A single configuration wavefunction Ψ depends on orthonormal functions ψ_i which are varied to minimize the energy

$$E = \langle\Psi\,|\hat{H}|\,\Psi\rangle. \tag{3.53}$$

This method provides a set of coupled differential equations for the ψ_i (Fock) with early applications by Hartree (HF theory).

In the early 1950s Roothaan treated molecular orbitals which were restricted to be linear combinations of a set of 3d 1-electron-functions χ_μ:

$$\psi_i = \sum_{\mu=1}^{N} c_{\mu i}\chi_\mu \tag{3.54}$$

Variation of the total energy was performed with respect to $c_{\mu i}$. We get a set of nonlinear equations providing a model, if the χ_μ are unambiguously determined by the nuclear positions. They are often denoted as self-consistent field (SCF) equations.

In the early days of MO theory, the χ_μ were chosen as the atomic orbitals building the molecule in linear combinations (therefore LCAO-SCF). The equations according to Roothaan can be extended to doubly/simply occupied orbitals and different spins on different orbitals. In the course of time, the challenge changed from numerical solutions of coupled differential equations (Hartree work) to evaluation of 3d or 6d integrals and solution of SCF equations. Best basic functions for LCAO-SCF methods were atomic orbitals of Slater type with exponential radial parts—in this case the one and two electron integrals can be solved analytically in the "two-centre case"—more centres were considered as a "nightmare" (John Pople) in the 1950s.

Two methods were used to treat these integrals: a semi-empirical one with a lot of empirical parameters and approximations using the PPP theory (Pariser, Parr, Pople, cf. Section 3.2.1.3). The other method tried to get an outcome without parameters ("ab initio") using Gaussian type basis sets. With the access to more powerful computers during the 1950s, theorists developed both type of programs (with Slater and Gaussian basis sets).

In 1968 the group of John A. Pople developed the famous GAUSSIAN program package. Unfortunately, the cost factor of calculating only small organic molecules was considerable. So JAP and Hehre developed a new integral algorithm that enhanced the calculation efficiency by a factor of >20. Moreover, they made a least-squares fit of Slater type basis functions by 3 Gaussians (STO-3G) and found good results for small molecules. This resulted in the package GAUSSIAN70. However, failures were found (e.g. at propene) and compensated by using two basis functions per valence atomic orbital (AO) instead of one (split-valence basis). Even with their sophisticated approaches, failures were observed—e.g., the ammonium molecule was predicted with a trigonal structure which was too flat, though. This difficulty could be overcome by mixing of higher (d) orbitals to the existing p-ones or regarding polarization effects. This Hartree–Fock model named HF-631G* was quite effective for many cases.

A general shortcoming of Hartree–Fock calculations, however, is the disregard of electron correlations between those of antiparallel spin. This leads to strong underestimation of bond dissociation energies, as being intrinsic to single determinant wavefunctions (WF). Hence, it is adequate to use linear combinations of HF determinants, stemming from unoccupied or virtual MO's—this leads to so-called spin orbital basis functions as products of the Cartesian basis functions and *a*/*b* spin functions. With label *i*, *j*, *k* for occupied spin orbitals and *a*, *b*, *c* for virtual ones, then a multi-determinant wavefunction may be written as

$$\Psi = a_0\Psi_0 + \sum_{ia} a_i^a \Psi_i^a + \sum_{ijab} a_{ij}^{ab}\Psi_{ij}^{ab} + \cdots \tag{3.55}$$

The *a*-coefficients can again be determined by variation to minimize the energy (configuration interaction CI), here in the case of double sets CID (doubles) or single ones included (CISD). Unfortunately, also these methods (CID and CISD) have limitations concerning large molecules/interacting systems. Regarding all possible substitutions we get the FID (full configuration model).

Electron correlation can also be taken into regard by perturbation theory:

$$\hat{H}(\lambda) = \hat{F}_0 + \lambda(\hat{H} - \hat{F}_0) \tag{3.56}$$

$\hat{F}_0$ is the Fock–Hamiltonian with the above Ψ_i being eigenfunctions. Then the energy can be expanded in powers of λ:

$$E(\lambda) = E_0 + \lambda E_1 + \lambda_2 E_2 + \lambda_3 E_3 + \cdots \tag{3.57}$$

This method was generated by Moeller and Plesset and developed to the power of 4 (MP1...MP4) giving more proper results in special cases—it was implemented in the GAUSSIAN system.

Another approach is the use of the following WF:

$$\Psi = (1 + T2)\,\Psi_0, \tag{3.58}$$

where T2 is an operator for all double substitutions leading to the coupled-cluster function CCD:

$$\Psi = \exp(T2)\, \Psi_0 \tag{3.59}$$

The CCD coefficients can be determined by a non-variational method and the procedure enhanced by simple substitutions leading to the CCSD method. This and the FID method provide good results for, e.g., composite two-electron systems like isolated He-atoms.

Summarizing the previous remarks, two quantities can be combined in one table, a two-dimensional model chart as given by JAP. The vertical column provides the different basis functions with increasing complexity starting from STO-3G to 6-311G (2df additions), the horizontal row the method of calculation from HF to FCI, also increasingly sophisticated with respect to electronic correlation. In those days of expensive computer power, calculations often were time-optimized, e.g., performing for raw predictions simpler basis sets and simpler models of correlation. In case of good results at this comparably low level, more sophisticated and longer calculations at the more difficult problems could be performed. The calculations mentioned so far provide reasonable energy differences, dissociation energies and ionization potentials for comparably small molecules. A common issue is the underestimation of binding energies, most probably due to the limited sets of basis functions (neglect of higher angular momentum basis sets). This problem can be overcome by introducing a small empirical correction (*sensu strictu* abandoning the ab initio principle), but the validation process comprises molecules up to benzene (42 electrons) with results being in the range of the target accuracy (see above). Besides some criticism of the methods mentioned above, JAP also left some comments on DFT methods, e.g. that two electron correlations are not treated exactly or that empirical parameters often have to be introduced. As we will see later, our DFT method does not suffer from these issues.

The procedures mentioned above last were state of the art in the 1970s to 1980s. It should be not forgotten that at those days computer power was extremely expensive ("CPU-time") as compared to nowadays where a simple smartphone disposes of a multiple performance of the early machines.

The author remembers his days of physics diploma, where a single run of the program on a Sperry Univac U1108™ needed several hours. It was even "worse" with the programs of the theorists.

3.2.1.2 The Nobel Prize winner Walter Kohn

Walter Kohn, born at March 9, 1923, in Vienna (Austria) was son of a middle class Jewish family earning their living on an art postcard edition publishing house. Having completed elementary school he attended a public high school in Vienna inner city. For about five years his interests mainly lay on the Latin and Greek language, not on mathematics or physics, as his later career may let suppose. The year 1938, with the taking over of Austria by Germany, changed everything: His father's business was confiscated without compensation, Walter Kohn was expelled from his school. Fortunately, he managed to enter a Jewish school with excellent teachers who awakened his high abilities in physics and mathematics. However, like his sister, WK had to emigrate to Gt. Britain. In the beginning he intended to become a farmer, but his foster parents fortunately managed to persuade him to attend a county school to continue his studies. However and strangely enough, he was interned by a general order of Prime Minister Churchill against "male enemy aliens": WK's formal German nationality was the crazy occasion. Having spent several months in different British camps, WK was finally shipped to Quebec City (Canada) and transferred to other camps there. Fortunately, WK received further skills in natural sciences by a special camp school and passed an university matriculation exam. He was then released from internment and went to Toronto, studying there physics and mathematics. At the end of one year's army service, he received a bachelor degree in the same field at Toronto university. From there he directly went to Harvard, to the famous theorist Julian Schwinger, and finished his thesis work (Kohn's variational principle of scattering) in 1948. For three years he stayed in the research group of Schwinger, then turned his interest from quantum field theory to solid-state physics and taught one of the first courses on this subject on behalf of famous J. H. Van Vleck. Being an expert in variational methods, WK developed a new variational approach to the study of wavefunctions in periodic crystals. In the frame of a National

Research Council fellowship he moved to Copenhagen and completely turned to solid-state physics as his main field of research. Having returned to Carnegie Institute of Technology in 1952, he mainly worked on a theory for the energy band structure of electrons in periodic crystals using his great knowledge in scattering, Green functions and variational methods (1950–1959). Meantime, he also worked at Bell Labs (H. Shockley, inventor of the transistor) and repeated his visits there till 1966. In the fall of 1963 WK spent a sabbatical semester at École Normale Supérieure in Paris and devoted his interest to the electronic structure of alloys. The question arose whether an alloy is completely or partially characterized by its electronic density distribution $n(r)$. There should be a close connection between a weak perturbing potential $\delta v(r)$ and the corresponding small change $\delta n(r)$ of the density distribution. For a single particle there also should exist an explicit elementary relation between $v(r)$ and $n(r)$ of the ground state. The density $n(r)$, according the assumption of WK, determines the external potential $v(r)$. The integral of $n(r)$ is the total number of electrons N—hence $n(r)$ would also determine the Hamiltonian H and hence all properties that can be derived from H and N, e.g. the wavefunctions of the 17th excited state Ψ_{17} $(r_1 \dots r_N)$, according to WK's own words. Then two different potentials $v_1(r)$ and $v_2(r)$ with different ground states Ψ_1 and Ψ_2 will produce the same density distribution. Together with the young researcher Pierre Hohenberg from USA, WK proved his assumption by help of the Rayleigh-Ritz variational principle, which was adapted from many-electron wavefunctions Ψ to the density $n(r)$—this is now known as the Hohenberg-Kohn variational theorem (HK). Starting from this, back in San Diego, WK developed together with his postdoc Lu J. Sham the famous Kohn–Sham (KS) equations, the basis for the present Density Functional Theory. In the following, WK moved to UCSB (University of California at Santa Barbara) to become initial director of the NSF Institute for Theoretical Physics (1979–1984), where he became Full Professor of Physics and Emeritus. However, he is still busy as a Research Professor. The climax of his career was the award of the Nobel Prize with John A. Pople in 1998—many other honours and prizes belong to his private Hall of Fame, among them numerous Honorary Doctorates.

3.2.1.3 The Nobel Prize winner John A. Pople†

John A. Pople, born at October 31, 1925, spent his early years in Burnham-on-Sea on the west coast of England. His father owned there a big clothing store, his mother's family were farmers, she herself worked as an army librarian during World War I. JAP attended Bristol Grammar School (BGS) rather far away from home, but his parents wanted to enable him to receive a good school education. In the age of 12 he detected his outstanding abilities in mathematics, later physics as well, which were extended by two excellent teachers at BGS. He was trained by them for a scholarship at Cambridge University, received an award for his excellent results and entered university in October 1943. During World War II, it was planned for students of high abilities to complete all degree courses and putting them to a government research institution afterwards. JAP did so, but, unfortunately with the end of the war neither government nor university had need for his services, as he wrote, due to the deluge of ex-servicemen of the British Army. Hence, JAP turned to an industry employment at Bristol Aeroplane Company. This employment did not satisfy him; he was not allowed to abandon the job, though, at these post-war days; so he applied for a return to Cambridge University, Trinity College, and succeeded in doing so in 1947. He attended many courses in almost any natural sciences and, finally, turned his interest to statistical mechanics of fluids, aside of the mainstream. In the following, he went to Sir Lennard Jones, an expert in molecular orbital theory (MO), and JAP subsequently became a theoretical chemist although his last experiences with chemistry were dating from the early 1940s at BGS. However, his efforts on the theory of hydrogen bonding were very fruitful, gained him a Ph.D. and a research fellowship at Cambridge. During the decade between 1950 and 1960 JAP tried to find a mathematical theory for simulating the whole chemistry. His efforts to apply theoretical models to practical properties led to the well-known PPP theory (Pariser, Parr, Pople).

At the end of 1955, JAP turned towards nuclear magnetic resonance, a very promising field of science in particular for chemical applications and spent two summers at National Research Council at Canada developing the theoretical background of NMR and comparing it with measured spectra. After an intermediate stay at National Physical Laboratory/London he spent a sabbatical year at Carnegie Institute of Technology in Pittsburgh (1962) and returned to there in 1964. At Pittsburgh he was attracted by fundamental problems of electronic structure and learned to generate computer codes for many of the most powerful machines at those days. In 1978 JAP and his group acquired the first VAX/780 minicomputer for use only at the chemistry department[7] and distributed programs to the general chemical community. He developed very sophisticated ab initio tools for the rather new field computer chemistry, modelling the wavefunctions from Slater or Gaussian orbitals. Legendary is his standard collection of programs "GAUSSIAN", the forerunner of other tools of the kind. In 1986 he moved to Northwestern University and continued his research programme at the chemistry department there. Strangely enough, JAP was excluded from further development and even the usage of his own (!) software. The reason allegedly was his refusal of the financial exploitation of scientific software. A hidden hint to his outstanding human character is the fact that he did not mention this incisive event in his biographical Nobel talk—let me mention this here as a personal remark. Fortunately, his merits were widely acknowledged by many honours, among them the appointment as the Knight Commander of the Order of the British Empire by Queen Elizabeth II in 2003.

Several journeys led him to universities at Australia, New Zealand, Israel and Germany in the 1980s and 1990s. The climax of his scientific career certainly was the award of the Nobel Prize for chemistry in 1998 together with Walter Kohn. John Pople died on March 15, 2004.

3.2.1.4 The self-consistent-charge Xα method (Michael Grodzicki)

In this chapter we will use much of the knowledge that was acquired above in Section 2.1. The most effective DFT method, where the physical properties of ions within their cluster surroundings of a well-known crystal structure are concerned, is the spin-polarized self-consistent charge (SCC)-Xα method in local spin-density approximation (LSDA) developed by Michael Grodzicki.

[7]In these early years of scientific computing, machines were installed and maintained at central locations and shared from many departments of a university.

This is in particular suitable for processing spectroscopic and diffractometric data, less for cases where the band structure of compounds plays the most important role.

The main thing here is the solution of the Kohn–Sham equations in order to find the one-electron density for the system involved. As we have shown above, it is defined as

$$\rho(\vec{r}) = \sum_k n_k \,|\psi_k(\vec{r})|^2 \tag{3.60}$$

with n_k being the occupation number of the k-th MO. The reasonability of the ψ_k is controlled by the relevant energy eigenvalues. The WF with the lowest energy will probably be the "best" one.

First, one has to construct a trial WF as the sum of a basis set connected to this WF. Near each nucleus it is reasonable to take, instead of molecular orbitals, atomic orbitals (AO) φ_i^I. Hence, it is also reasonable to construct an approximate MO-WF as a linear combination of AO's (LCAO or Hartree–Fock (HF)-Roothaan approximation):

$$\psi_k(\vec{r}) = \sum_{I,i} c_{ik}^I \varphi_i^I(\vec{r} - \vec{R}_I) \tag{3.61}$$

where $\{\varphi_i^I\}$ (basis set) is the i-th AO of the I-th nucleus and c_{ik}^I the expansion coefficient of this AO orbital to the MO ψ_k. The choice of the most suitable basis set is decisive for the further course of calculation.

Three types of basis sets are commonly used:

(1) All-electron: all fully or partially occupied AO's in the atomic ground-state are included (e.g. 1s, 2s, 2p for O)
(2) Valence: Only the relevant orbitals are taken (e.g. 2s, 2p for O)
(3) Extended: Extra orbitals are added (e.g. non-occupied d-orbitals)

Then, using the MO-WF ψ_k thus constructed, we may write the Schrödinger equation as

$$\hat{H}\psi_k(\vec{r}) = \varepsilon_k \psi_k(\vec{r}) \tag{3.62}$$

We then have to solve the following secular equation:

$$\sum_{I,i} (H_{ij}^{IJ} - \varepsilon_k S_{ij}^{IJ}) c_{ik}^I = 0 \tag{3.63}$$

The matrix elements of H_{ij}^{IJ} are constructed as integrals between $\varphi_j^J(\vec{r}-\vec{R}_J)$ and $\varphi_i^I(\vec{r}-\vec{R}_I)$ using as elements a kinetic term and a molecular part V_{mol} comprising Coulomb/Exchange potential $V_C[\rho(r)] + V_{xc}[\rho(r)]$. S_{ij}^{IJ} is the overlap integral containing only the above WF's describing two orbitals of interacting atoms. The value of this overlap integral is very important and describes the strength of bonding interactions, i.e. strong bonds are due to strong overlap.

How "good" a DFT method describes the physical reality is decided by the accuracy of the calculations of the above integrals. Naturally, this is warranted by a 3D-numerical integration. However, using the SCC-Xα approximation, the calculations are reduced by (allowed) model simplifications and confinement to one-dimensional numerical integrations:

First, an atomic Hamiltonian is defined:

$$H_{\text{at}}^I = -\frac{\hbar^2}{2m_e}\nabla^2 + V_{\text{at}}^I(\vec{r}-\vec{R}_I) \tag{3.64}$$

with atom/nucleus at site R_I. Now the above Hamiltonian H_{ij}^{IJ} is split into Hamiltonians for atoms I and J constructed according to the last equation and giving instead of V_{mol} a neighbour potential V_{nb} containing a difference between V_{at}^I and V_{at}^J. On the right side of the Schrödinger equation we get two energies ε_i^I and ε_i^J, the valence-state ionization potentials of atoms I and J with, e.g., for atom I:

$$\varepsilon_i^I = \varepsilon_{i0}^I + \varepsilon_{i1}^I Q_I + \varepsilon_{i2}^I Q_I^2 \tag{3.65}$$

This is just a power development in the effective charge Q_I containing a core charge part Z_I^c and the number of valence electrons:

$$Q_I = Z_I^c - N_I^{\text{val}} \tag{3.66}$$

The core charge may be, e.g., +4 for Si. The main goal is now, to find a simple model potential which nevertheless is able to describe the crystal structure and the physical properties properly.

Second, the molecular electron density is developed as the sum of atomic densities:

$$\rho(\vec{r}) = \sum_K \rho_{\text{at}}^K(|\vec{r}-\vec{R}_K|) \tag{3.67}$$

Also, the molecular potential can be approximated as a sum of K atomic Coulomb potentials:

$$\hat{V}_{\text{mol}}(\vec{r}) = \sum_{K} \hat{V}_c^K(|\vec{r} - \vec{R}_K|) + \hat{V}_{\text{xc}}[\rho(\vec{r})] \tag{3.68}$$

We now refer everything to the I-th atom and define the difference vector

$$\vec{R}_{JI} = \vec{R}_J - \vec{R}_I \tag{3.69}$$

So the Hamiltonian matrix elements will give:

$$H_{ij}^{IJ} = \int \varphi_j^J(\vec{r} - \vec{R}_{JI}) \left\{ H_{\text{at}}^I + \sum_{K \neq I} \hat{V}_C^K(|\vec{r} - \vec{R}_{KI}|) + \hat{V}_{\text{xc}}[\rho(\vec{r})] - \hat{V}_{\text{xc}}[\rho_{\text{at}}^I(\vec{r})] \right\} \varphi_i^I(\vec{r}) d^3r \tag{3.70}$$

From experience it is known that the one-centre integrals, e.g.

$$\int \varphi_j^J(\vec{r} - \vec{R}_{JI}) H_{\text{at}}^I \varphi_i^I(\vec{r}) d^3r, \quad I = J \tag{3.71}$$

are larger by almost an order of magnitude than the two- and three-centre integrals with, e.g., $K \neq I \neq J$.

Table 3.1 Types of integrals in Hamiltonian matrix elements

$I = J$	$\int \varphi_j^I(\vec{r}) H_{\text{at}}^{II} \varphi_i^I(\vec{r}) d^3r$	1-center atomic
$K \neq I = J$	$\int \varphi_j^I(\vec{r}) v_{cb}^K(|\vec{r} - \vec{R}_{KI}|) \varphi_i^I(\vec{r}) d^3r$	Coulomb crystal-field
	$\int \varphi_j^I(\vec{r}) v_{xc}[\rho(\vec{r})] \varphi_i^I(\vec{r}) d^3r$	exchange crystal-field
$I \neq J$	$\int \varphi_j^I(\vec{r} - \vec{R}_{JI}) \varphi_i^I(\vec{r}) d^3r$	overlap
$I \neq J$	$\int \varphi_j^J(\vec{r} - \vec{R}_{JI}) H_{\text{at}}^{IJ} \varphi_i^I(\vec{r}) d^3r$	resonance
$K \neq I = J$	$\int \varphi_j^J(\vec{r} - \vec{R}_{JI}) v_{\text{cb}}^I(\vec{r}) \varphi_i^I(\vec{r}) d^3r$	2-center Coulomb
$K \neq I = J$	$\int \varphi_j^J(\vec{r} - \vec{R}_{JI}) v_{\text{cb}}^K(|\vec{r} - \vec{R}_{KI}|) \varphi_i^I(\vec{r}) d^3r$	3-center Coulomb
$I \neq J$	$\int \varphi_j^J(\vec{r} - \vec{R}_{JI}) v_{\text{xc}}[\rho(\vec{r})] \varphi_i^I(\vec{r}) d^3r$	2-and 3-center exchange

Hence, it is necessary to calculate the former much more exactly than the latter, because failures in the former would dominate the results by far. So the main merit of the SCC-Xα approximation is to treat one-centre and multi-centre integrals with different accuracy. A table of all occurring integrals is given in the previous table (Table 3.1) according to Zherebetskyy (2010).

3.2.1.5 The evaluation of the multi-centre integrals

Contrary to the one-centre integrals we may calculate the multi-centre ones with minimal basis sets, simple core potentials and the local exchange-only functional Xα (the correlation term is neglected). Furthermore, the atomic WF can be divided into a radial and into an angular part, as known from elementary quantum mechanics:

$$\varphi_{nlm}(\vec{r}) = R_{nl}(r)Y_{lm}(\theta,\phi) \tag{3.72}$$

with n, l, m being the quantum numbers characterizing main, orbital and magnetic atomic orbitals, whereas $Y_{lm}(\theta, \varphi)$ can be defined over the famous Legendre polynomials giving the widespread orbital images in textbooks. Some of them can be also perceived in the present book (cf. Section 3.3.3), although most of them are hybrid orbitals. The radial parts R_{nl} describe the extension of the relevant atoms and one may take hydrogen atom radial functions for this purpose, but integration is difficult; so a simpler point of departure would be to take the radial part of Slater-type orbitals (STO's):

$$R_{nl}(r) = a_{nl}e^{-\zeta_{nl}r} \tag{3.73}$$

with a_{nl} being a normalization constant and ζ_{nl} an orbital and charge dependent parameter:

$$\zeta_{nl} = \zeta_{nl,0} + \zeta_{nl,1}Q \tag{3.74}$$

This concerns the strength of the bond: The stronger an electron is bound in the atom, the more localized is the corresponding WF for s and p orbitals.

Each valence orbital can now be expressed by help of a radial STO and a spherical harmonic function:

$$\varphi_{nlm}(\vec{r}) = a_{nl}r^{n-1}e^{-\zeta_{nl}r}Y_{lm}(\theta,\phi) \tag{3.75}$$

Another approximation concerns the atomic particle density:

$$\rho(\vec{r}) = \sum_i X_i e^{-\eta_i r} \tag{3.76}$$

with η_i being a numerical charge-dependent parameter:

$$\eta_i = \eta_{i0} + \eta_{i1}Q \tag{3.77}$$

Furthermore, the particle density can be separated into a core part and into a valence part:

$$\rho(\vec{r}) = \sum_K \rho^K_{\text{core}}(|\vec{r} - \vec{R}_K|) + \sum_K \rho^K_{\text{val}}(|\vec{r} - \vec{R}_K|) \tag{3.78}$$

With the Xα approximation, it is assumed that the exchange-correlation potential can be written as

$$\hat{V}_{\text{xc}}[\rho(\vec{r})] = -3\alpha\left(\frac{3}{\pi}\right)^{1/3}\left\{\sqrt[3]{\rho_{\text{core}}(\vec{r})} + \sqrt[3]{\rho_{\text{val}}(\vec{r})}\right\} \tag{3.79}$$

A principal assumption is that the densities might be described in the same form:

$$\rho^K(\vec{r}) = \frac{N_K\eta_K^3}{8\pi}e^{-\eta_K|\vec{r}-\vec{R}_K|} \tag{3.80}$$

with the above charge dependent valence density. We may then write the Coulomb potential $V_C{}^K$ by help of radial amounts and spherical harmonics. A further advantage of the above expression is the evaluation of the molecular exchange potential:

$$V^I_{\text{xc}}[\rho(\vec{r})] = -\alpha'\left(\sum_K N_K\eta_K^3 e^{-\eta_K|\vec{r}-\vec{R}_K|}\right)^{1/3} \tag{3.81}$$

with α' being a constant factor (see V_{xc} above) times α.

Also this exchange potential can be developed with spherical harmonics in a one-centre expansion formula. Hence, V^I_{xc} from the neighbours and the Coulomb potential exhibit the same angular dependence; hence, one may express the molecular potential from the neighbours as one-centre series-expansions with respect to R_I. Moreover, the l-summation may be terminated at a value of 6 since bigger shells are not realistic. Summarizing, we have several quantities depending on the effective charges of the atoms like $\varepsilon_{nl}{}^I$, ζ_{nl}, η_i with

$$Q_I = Z_I^{\text{core}} - N_I^{\text{val}} \tag{3.82}$$

The number of the valence electrons N_I^{val} is calculated from the overlap population analysis of Mulliken (1955) for cluster orbitals:

$$N_{ij}^{IJ} = P_{ij}^{IJ} \cdot S_{ij}^{IJ} \tag{3.83}$$

S_{ij}^{IJ} is the already known overlap integral between two AO's and P_{ij}^{IJ} the bond-order matrix describing the electron distribution on the basis set AO's.

The number of valence electrons at the I-th atom is written as

$$N_I^{\text{val}} = \sum_i N_i^I \tag{3.84}$$

It remains to ask, how in the frame of SCC-Xα calculations we can know, when to break up. The criterion is the difference between the orbital occupation numbers between two subsequent iterations n and $n + 1$ falling below a given accuracy limit, so that

$$\max_{I,i} \left| N_i^{I,(n+1)} - N_i^{I,(n)} \right| < \varepsilon \tag{3.85}$$

Summarizing, we mention the steps for the self-consistent solution of the HF-Slater equation (SCC-Xα approximation):

(a) Initial guess of suitable orbital occupation numbers or expansion coefficients for AO basis sets
(b) Solving of the secular equation:

$$\left| H_{ij}^{IJ(0)} - \varepsilon_K^{(0)} S_{ij}^{IJ(0)} \right| = 0 \tag{3.86}$$

(c) the eigenvalues $\varepsilon_K^{(0)}$ from the Schrödinger equation are calculated.
(d) By iteration we take the eigenvalues $\varepsilon_K^{(0)}$ and a new set of expansion coefficients (orbital numbers) as point of departure and calculate improved elements $H_{ij}^{IJ(1)}$ and $S_{ij}^{IJ(1)}$. Afterwards $\varepsilon_K^{(1)}$ are determined, and so forth.
(e) This iterative procedure should continue till a given threshold as mentioned above will be exceeded.

By the above procedure all symmetry properties of the system under consideration are preserved; CPU time and memory space by simplifications in the integration are reduced. The results of the calculation are reliable in spite of approximations and simplifications—as proved in many relevant cases such as the one described in the present book. A slight shortcoming is the dependence on very good structural data, but these are available for most of important samples. Hence, the procedure is nicely applicable for real crystallographic samples and other large and complex systems.

3.2.1.6 The Program WIEN2k by Peter Blaha and Karlheinz Schwarz

Not every scientific group is lucky enough to have a theoretical physicist like M. Grodzicki among the own ranks to perform DFT calculations. For those colleagues, the famous Vienna theoretical physics group (Blaha et al., 2014) has created a generally accessible program package named WIEN2k. It consists of several FORTRAN90 routines connected by C-shell scripts—it is platform-independent and can be run using web browsers, a special interface or specialized commands. It allows to calculate electronic structures of solids using the DFT method and is based on the full-potential (linearized) augmented plane wave (LAPW) + local orbitals (lo) method and is as such most suited to band-structure calculations. Being an all-electron method, it also regards relativistic effects and is used by estimated 2000 user groups worldwide with a specialized workshop every year at different locations in the World.

Main applications of the method are

(1) energy bands and density of states
(2) electron spin densities, X-ray structure factors
(3) total energy, forces, equilibrium geometries, structure optimization, molecular dynamics
(4) phonons
(5) efg, δ's, $H(0)$'s
(6) spin-polarization (ferro- or antiferromagnetic structures), LS coupling
(7) X-ray emission/absorption spectra, Electron Energy Loss Spectroscopy EELS
(8) optical properties
(9) Fermi surfaces
(10) centro/non-centrosymmetric cells (230 SG's included)

The common property of the method presented above and the program WIEN2k is the underlying basis, the Density Functional Theory (DFT). As pointed out by Blaha (2010), this method might be considered as an exact one, but the underlying functional is unknown so that the user is forced to make approximations. For the method applied by us, it was the LSDA (other possibility is GGA, generalized gradient approximation) where the local electron density enters the E_{xc} (exchange correlation energy) and the relevant potential with good success in calculating the efg and other physical quantities. However, in systems with strongly correlated electrons in 3d transition metal oxides or 4f/5f compounds the above approximation may fail and it has to be extended to more sophisticated calculations like LDA + U a.s.o.

In particular, with regard to the determination of Mössbauer parameters, where we have to consider a possibly accurate description near the nucleus, plane wave methods often show to be inadequate. Therefore an extension for the latter is used, the augmented plane wave (APW) based methods like the program mentioned in this subchapter. In principle, the unit cell is divided into two regions, one with non-overlapping atomic spheres described by a linear combination of numerical radial functions $u_l(r, E)$ times spherical harmonics $Y_{lm}(r)$:

$$\phi_{kn} = \sum_{lm} A_{lm} u_l(r,E) Y_{lm}(r) \tag{3.87}$$

The other one is an interstitial region, where a plane wave expansion can be applied:

$$\phi_{kn} = \frac{1}{\sqrt{\omega}} e^{ik_n r} \tag{3.88}$$

with ω being the unit cell volume and $k_n = k + K_n$ the wave vectors outside the first Brillouin zone characterized by k. u_l is a regular solution of the radial part of the Schrödinger equation, which leads to a non-linear eigenvalue problem, but can be linearized, e.g. by additional local orbitals (APW + lo method, like in WIEN2k). The coefficients are chosen in such a way that the additional basis function attains zero for the boundary of the sphere. In WIEN2k the convergence of the basis set is controlled by a cut-off parameter—the method was successfully tested at unit cells with more than 1000 atoms, which is a comparably large quantum mechanical system.

Within the frame of WIEN2k, among other parameters, an efg can be calculated from the electronic and nuclear charge distribution by

$$V_{zz} = \int \rho(r)\frac{2P_2(\cos\theta)}{r^3}dr \tag{3.89}$$

with P_2 being second order Legendre polynomials. The density ρ can be decomposed into contributions from 3p, 4p, 3d orbitals—as pointed out by Blaha (2010), the efg "can be very sensitive to structural details...", this is fully supported by our findings although we have used quite another program system.

Therefore, considering that WIEN2k gives an easy access to Mössbauer parameters and is created for the broad scientific community, we can warmly recommend it, also for comparison with the results of our DEN method presented here. It would be highly desirable to test the program at our specific sample which has not already been done so far.

3.3 The Semi-Quantitative Approach to Obtain an efg from Diffractometer Data

3.3.1 Fundamentals of Diffractometry

The term "diffractometry" means the fraction or scattering of incoming waves at an inflecting structure. Till the end of the 19th century this term was exclusively understood in the connection of "fraction of light". The famous experiment of Isaac Newton who first observed the splitting of sunlight into its coloured components is an example for the phenomenon "fraction". However, diffraction is more: It also includes the well-known observation that light waves are scattered at a narrow slit. In particular, if we use a double slit, it is possible to see strange variations of bright and dark regions on a screen behind the slits. If we increase the number of slits, these regions will become more and more numerous and distinct. In principle, the effect can nicely be verified by drawing two sine functions on a sheet of paper with a phase difference of $\pi/2$ or $\lambda/4$ and subsequently adding the two waves.

The sum wave will receive an increased height (as compared to its single components) and a phase shift with respect to both. If the two waves have a phase difference of π (or $\lambda/2$), we will observe a zero line since the two waves annihilate each other. If the phase difference, however, is zero, the two waves will add up to a sum wave which is twice as high in amplitude than the single components.

These effects are called interference, or in detail, destructive interference for the second case mentioned above (complete annihilation) and constructive interference for the first one (complete amplification). Which are the conditions for interference of waves?

(1) The interfering waves must be coherent i.e. both waves must stem from the same (point) source—a flashlight is not suitable for such experiments.

(2) The inflecting smallest structural units must be comparable with the wavelength of the incident waves.

Nowadays, such experiments can be nicely repeated by help of a LASER pointer, e.g. let the beam hit a suitable CD surface under a grazing angle (or better: the removed lid of a common dynamic photo frame) and observe the nice interference spots on a screen behind the structure.

After the discovery of X-rays in the late 19th century by Wilhelm Conrad Röntgen, who was awarded the first Nobel Prize of Physics for this finding, Max von Laue has had the genius idea that the assumed regular arrangement of atoms in a crystal structure might form a scattering lattice for the recently detected X-rays, similarly as the light waves had behaved at the regular system of slits or points at the time of Huygens. X-rays are able to interfere (since they stem from a comparably small source) and their wavelength amounts to the order of 1 Å [= 0.1 nm], indeed the size of single atoms or ions and the distance between them. Indeed, von Laue exposed a crystal to the X-rays of the rather newly disposable tube and found interference spots on a photographic plate behind the crystal. Moreover, he explained why this should be occurring by explaining the mechanism of the X-ray scattering and subsequent interference of the outgoing waves.

The scattering is produced by the incident X-rays hitting the electronic shells of the crystal atoms or ions. Nowadays, the term "shell" (stemming from Niels Bohr) is old-fashioned but it still serves its purpose for the explanation of special phenomena. The electronic shells will then be excited to forced oscillations and the generation of secondary waves which interfere with each other and dispose of distinct properties:

(1) The direction of these waves is determined by the shape and size of the crystal unit cell involved (lattice constants, angles, crystallographic system).
(2) The intensities of the interference spots on the photographic plates/films are determined by the contents of the unit cell—the position and type of atoms or ions within the cell.

It was a sophisticated extension of the former Fraunhofer-Fresnel theory to a 3D crystallographic structure as the scattering lattice. The interference of the diffracted waves was explained by the famous Laue equations which shall be derived in the following:

Imagine two crystallographic unit cells in the distance of a translation vector $\mathbf{t} = t_1\mathbf{a} + t_2\mathbf{b} + t_3\mathbf{c}$, where **a**, **b**, **c** are lattice vectors and t_1, t_2, t_3 the whole numbers. Let us further assume an incident X-ray beam with the direction of a unit vector $\mathbf{s}_0$ and a scattered beam with the unit vector **s**, diffracted under an angle θ (Fig. 3.25).

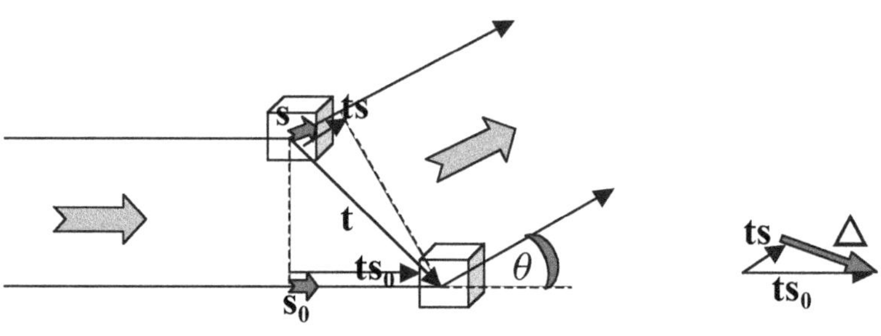

Figure 3.25 Scattering of incoming X-rays at two neighbouring unit cells connected by a translation vector **t**, schematically, according to *v*. Laue. Unit vector in the direction of the primary beam denoted by $\mathbf{s}_0$, the one of the scattered beam (reflection), by **s**; both vectors are correlated by the diffraction angle θ. Small insert at the right hand side describes the path difference Δ of the waves.

The two scattered waves should undergo constructive interference and produce a reflection on a given screen. Then the path difference between the two outgoing waves can be given as (Fig. 3.25)

$$\Delta = \mathbf{ts}_0 - \mathbf{ts} = \mathbf{t}(\mathbf{s}_0 - \mathbf{s}). \tag{3.90}$$

With the above translation vector **t** we then obtain

$$\Delta = t_1\mathbf{a}(\mathbf{s}_0 - \mathbf{s}) + t_2\mathbf{b}(\mathbf{s}_0 - \mathbf{s}) + t_3\mathbf{c}(\mathbf{s}_0 - \mathbf{s}) \tag{3.91}$$

Now we may demand that the path difference Δ be for each **t** (for each triple $t_1t_2t_3$) a whole multiple of the wavelength λ. In terms:

$$\Delta = t_1h_1\lambda + t_2h_2\lambda + t_3h_3\lambda, \tag{3.92}$$

where h_1, h_2, h_3 are also whole numbers.

The reason is that otherwise we would not obtain constructive interference between the outgoing waves.

Comparison of the two above equations will give three equations for each lattice vector, namely:

$$\mathbf{a}(\mathbf{s}_0 - \mathbf{s}) = h_1\lambda;\ \mathbf{b}(\mathbf{s}_0 - \mathbf{s}) = h_2\lambda;\ \mathbf{c}(\mathbf{s}_0 - \mathbf{s}) = h_3\lambda\ h_1, h_2, h_3 \in Z \tag{3.93}$$

Now we may resolve the products of the lattice constants with the unit vectors, e.g.

$$\mathbf{a}(\mathbf{s}_0 - \mathbf{s}) = \mathbf{as}_0 - \mathbf{as} = \cos(\angle\mathbf{a}, \mathbf{s}_0)\ |\mathbf{a}||\mathbf{s}_0| - \cos(\angle\mathbf{a}, \mathbf{s})\ |\mathbf{a}||\mathbf{s}| \tag{3.94}$$

and analogously for the other two equations. If we define the angle between **a** and $\mathbf{s}_0$ as α_0 and the angle between **a** and **s** as α (analogously for the other relevant angles), we will obtain

$$\begin{aligned}(\cos\alpha_0 - \cos\alpha) &= h_1\lambda/\alpha \\ (\cos\beta_0 - \cos\beta) &= h_2\lambda/b \\ (\cos\gamma_0 - \cos\gamma) &= h_3\lambda/c.\end{aligned} \tag{3.95}$$

These are the famous Laue equations describing the interference conditions for any scattering waves at a crystal lattice. Obviously, this is an overdetermined system of equations. In order to find solutions, we have to fix one parameter and release the other and vice versa, namely:

(1) λ is fixed and $\mathbf{s}_0$ is variable.

This is equivalent to a monochromatic X-ray beam (X-rays with a uniform wavelength) and the direction of the incident (primary) beam variable.

This corresponds to an experiment with an oscillated crystal and an incident monochromatic beam; in practice this is verified in the so-called Bragg-turning-crystal method.

Another procedure being equivalent to the latter would be to transform the crystal to a powder. In this arrangement, the monochromatic beam will hit tiny crystallites which will expose—by their locally different positions—also different scattering angles to the beam. We then do not get simple reflections any more but whole rings of undistinguishable reflections of crystallites lying rotational symmetrically with respect to the primary beam. The incident beam is scattered into secondary rays forming different cone mantles. We then record the cutting lines of the cone mantles on a photographic film/plate—this procedure is called Debye–Scherrer method.

The second general possibility is

(2) λ is variable and $\mathbf{s}_0$ is fixed.

This is the constellation of the famous *experimentum crucis* which was performed by von Laue and co-workers:

The wavelength of the X-rays be not uniform but polychromatic—for this purpose, we need a fixed crystal with respect to the beam direction. At this procedure, only those reflections are produced by constructive interference whose wavelengths will fulfil the requirements of the Laue equations. It is difficult to obtain lattice constants from the array of the reflections obtained, but good results concerning the reduced crystallographic symmetry can be well established, the so-called Laue symmetry or 11 Laue classes (out of 32 possible point symmetries in a lattice). Why not more than that? The reason is that this Laue method cannot distinguish whether a beam scattered at a centrosymmetric crystal be diffracted at the front face of the crystal or the inversely connected back face. So crystals containing a centre of inversion symmetry cannot be distinguished unambiguously by the Laue method—therefore the reduction of the crystal classes.

Nevertheless, these findings were outstanding in those early days, and von Laue received the Nobel Prize of Physics in 1914 for his discovery. In this sense he should be considered as the father of crystal diffractometry, but there are many "sons" and "daughters" dedicating their life's work to this fascinating topic.

A very convenient form to determine the metric properties of a unit cell, however, is given by Sir William Lawrence and Sir William Henry Bragg (Nobel Prize for father and son (sic!) in 1925).

They described the lattice planes of a crystal as almost ideally reflecting mirrors for X-rays and the derivation of their formula for crystal scattering is stunningly straightforward (Fig. 3.26).

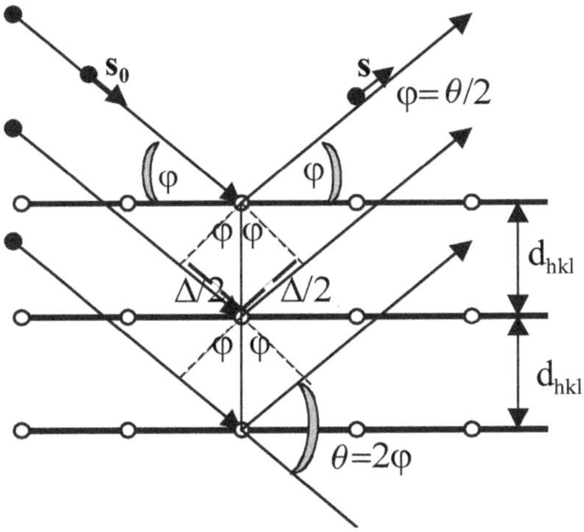

Figure 3.26 Reflection and interference of incoming X-rays at neighbouring crystal net planes in the distance d_{hkl} under a glancing angle φ, schematically, according to W. H. and W. L. Bragg. Unit vector in the direction of the primary beam is denoted by $\mathbf{s_o}$, the one of the scattered beam (reflection), by $\mathbf{s}$. The path difference Δ between the two waves leading to the famous Bragg equation can nicely be deduced from trigonometric considerations.

Let us assume different X-rays of the primary beam hitting crystal planes under the glancing angle φ. This multitude of rays will be reflected at parallel stacks of net planes characterized by the same *hkl*-triple with an internal distance d_{hkl}. From Fig. 3.26 it may be noticed that two neighbouring rays have different path lengths, the lower one receiving an additional path length which depends on the distance between the lattice planes, the sine of glancing angle φ (being half of the scattering angle θ, cf. Fig. 3.26) and the wavelength. According to the drawing, two times the sine of half of the scattering angle times the distance between neighbouring *hkl* planes gives the wavelength λ or a whole multiple of it. In terms:

$$2d \text{ sine } \theta/2 = m\lambda \quad (3.96)$$

This is the famous Bragg equation describing the condition of scattering and constructive interference, equivalent to the Laue equations. Indeed, both formulae could be transformed into one another, as it should be; see, e.g., my online book *Festkörperphysik* (Lottermoser, 2015).

The Bragg equation can be easily interpreted in terms of the different methods mentioned above:

3.3.1.1 Bragg procedure

A monochromatic X-radiation may hit a crystal. The latter and its planes will be turned till the glancing angle will for each plane fulfil the Bragg condition: We will get an array of reflections (belonging to each set of net planes) which can be indexed with triples of *h*, *k*, *l*. So each net plane (or parallel set of net planes) of a crystal corresponds to a reflection on a fluorescent screen or photographic plate/film. Unfortunately, we do not see the real lattice through this array of points, but the so-called "reciprocal" one. This is due to the reflection phenomenon—below we will also come back to the construction of the reciprocal lattice. In any case, by the Bragg procedure it is possible to find the distance between net planes or even lattice constants by rewriting of the Bragg equation, e.g. for orthogonal lattices:

$$d_{hkl} = \lambda/\sqrt{(h/a)^2 + (k/b)^2 + (l/c)^2} \quad (3.97)$$

3.3.1.2 Powder method (Debye–Scherrer)

As already mentioned above, we also need a monochromatic X-radiation; the turning of one crystal and subsequent recording of relevant reflections in case of interference according to Bragg's law will be replaced by producing lots of crystallites ("the powder sample") and letting them "bath in the beam". The rotational symmetric rings (whose origin was discussed above) can now be recorded on a stripe of photographic film being attached to the inner wall of a cylinder whose axis is identical to that of the small stage where the powder sample is located. Unrolling the film will give more or less curved lines (parts of the Debye–Scherrer-rings) with different thickness i.e. intensities of the scattered rays. From the distance of these lines with respect to the point where the primary beam hits the film, we may get the different scattering angles. From these it is possible—using Bragg's law—to calculate the *d*-values of the different lines or "reflections", respectively, Together with the intensities of the "reflections" and the indexing of them, the researcher will obtain everything to determine at least the lattice constants of the sample involved. This is classical powder diffractometry. Later on the cylinder with the film stripe was replaced with an electromechanical driven lever with a counter tube, by which the intensity of the reflection lines and their associated angle values could be determined quantitatively. The plate with the crystallite powder and the lever had then to be turned in a correlated motion, providing the utmost possible number of Bragg conditions i.e. reflection lines. In Fig. 3.27 a typical powder spectrum is shown:

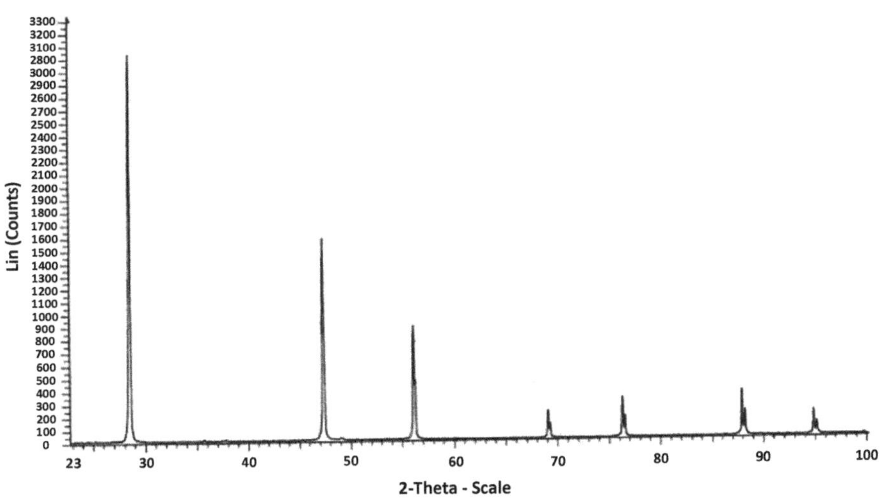

Figure 3.27 X-ray powder diffraction spectrum of silicon: Count rate of γ-quants (intensity of the reflections) is dependent on double scattering angle. The spectrum can be considered as a "finger-print" of the involved sample helping to identify the occurring mineral or mineral phases.

There are several tools to optimize the devices constructed for these purposes by introducing Soller slits, monochromator crystals, different geometries for the arrangement of the sample, primary beam and lever, but we will stop at this point since powder diffractometry was not applied for our purposes. However, it may be said finally that nowadays it has become possible to refine the different Bragg peaks by certain profiles in order to not only get lattice constants with very high accuracy but also make structure refinement which was deemed impossible in the early days of diffractometry ("Rietveld refinement"). The main purpose of powder diffractometry, however, is the determination of crystal phases within unknown powder mixtures.

3.3.1.3 Laue procedure

As mentioned before, this technique was used by the discoverer of the X-ray diffractometry himself and therefore bears his name. This method has the great advantage that the experimental assembly is comparably simple—the disadvantage is that the evaluation of lattice constants or indexing is rather complicated due to the non-linearity of the reciprocal lattice lines marked by the reflections on the screen or photographic film. Contrary to the Bragg methods, the incident radiation does not need to be monochromatized. This means that the equipment lacks of filters or monochromator crystals (see Section 3.3.2) since the polychromatic X-radiation "fresh from the X-ray tube window" can be used. Therefore, without further assemblies besides a collimator to render the beam sufficiently parallel, a polychromatic beam hits the crystal under consideration. From the different crystal planes only reflections are concerned which fulfil with a given glancing angle and d_{hkl} of the relevant net planes the Bragg condition. We might imagine to get reflections of "different colours", since the differently tilted lattice planes select the incoming primary beam according to the "colour" of the ray. It is evident that we have no sensor organ for distinguishing the different wavelengths of an X-ray beam, but we might compare the phenomenon with the dispersion of white light in a prism. Unfortunately, the reflections of different colours and different net planes can also coincide, so it is difficult to analyse a Laue diffractogram. Moreover, the different outgoing rays are scattered into tilted cone mantles. The projection onto a given plane behind the crystal does not produce reflections on straight lines but on cone cuts which may be strangely tilted according to the orientation of the relevant net planes. Therefore, it is preferred to have the crystal (by chance or by optical, morphological etc. analysis) with a lattice vector arranged parallel to the primary beam. In this case we often observe highly symmetrical, very impressing patterns of reflections showing the crystal symmetry in this direction. For example, having the principal six-fold axis of a crystal with hexagonal structure parallel to the primary beam, we will observe this six-fold symmetry on the screen or film.

Like this, we may also observe mirror planes, four-fold axes, a.s.o.—almost everything besides inversion centres. As already mentioned above, it is not possible to uniquely identify crystals containing an inversion centre. A nice example for a highly symmetric Laue pattern is given in Fig. 3.28:

With a general direction of the crystal being parallel to the primary beam the cone cuts will be distorted up to displaying no symmetry at all. In this case, it is not easy to find a lattice vector or to perform indexing by attributing values of h, k, l (with h, k, l being whole numbers) to the occurring reflections. In order to understand the main properties of diffractometry better, several words on the theoretical background of the method should be left here.

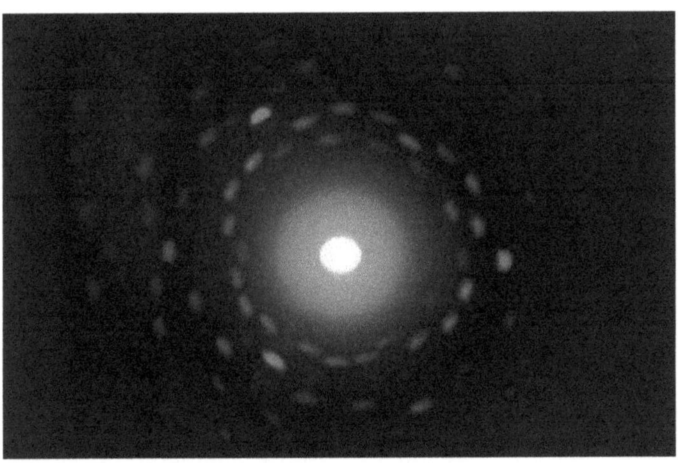

Figure 3.28 Laue pattern of a quartz crystal plane with its trigonal axis parallel to the X-ray beam (radiation is Mo with $\lambda = 0.7107$ Å).

3.3.1.4 Theoretical background of some crystallographic properties

Parallel to or even some time before von Laue's outstanding findings, basic concepts of the structure of crystals and the theory of diffraction have been worked out.

First, it should be explained what is understood by the unit cell of a crystal. The most general shape of a unit cell is mathematically a so-called parallelepiped. This is a body with 6 limiting planes all in the shape of a parallelogram. The basis vectors of these planes are established by a tripod with three different angles different from 90° and three different lengths, so that

$$|a| \neq |b| \neq |c| \text{ and } a \neq b \neq \gamma \neq 90° \tag{3.98}$$

An illustration of the properties is given in Fig. 3.29.

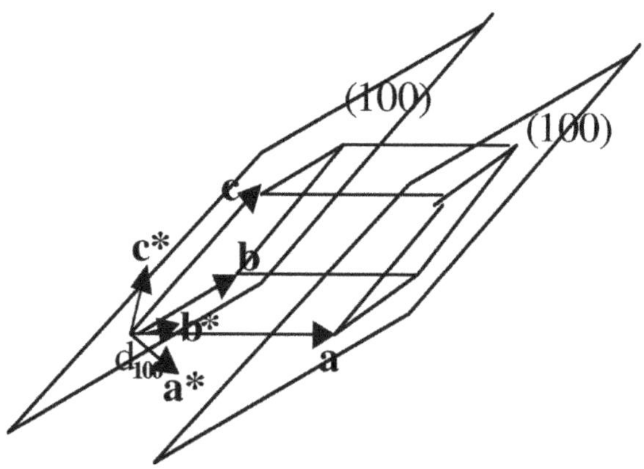

Figure 3.29 General shape of a unit cell in the triclinic case (all lattice constants a_0, b_0, c_0 being different from each other, all angles being unequal to 90°). Mathematically, it is called "parallelepiped". Two parallel net planes are emphasized (100), the reciprocal lattice vectors **a***, **b***, **c*** are perpendicular to the principal planes. The length of, e.g., the lattice vector **a*** corresponds to the distance d_{100}.

Let us consider the planes at the side faces of the parallelepiped. According to a common notation which has been introduced earlier, these planes are characterized by their inclination in space and their intersection with the crystallographic axes **a**, **b**, **c**. The plane on the right hand side of the above body intersects the *a*-axis in one unit, the other axes in infinity. So it would be possible to characterize the mentioned plane by 1, ∞, ∞. The term infinity, however, is somehow vague. So crystallographers decided to take the inverse of it: 1, 0, 0 to avoid this difficulty. For other inclinations, the procedure of this "indexing" is more arduous. For example, if the three crystal axes are intersected in $2a$-units, $3b$-units and 4 c-units, then we have to take the inverse $\frac{1}{2}:\frac{1}{3}:\frac{1}{4}$ and find the common denominator: 2 × 3 × 4 = 12. Hence, we get $\frac{6}{12}:\frac{4}{12}:\frac{3}{12}$. Now we take the numerator and get the so-called Miller indices h, k, l of this plane: (643).

The unit cell of the above-mentioned parallelepiped is limited by two parallel planes with the indices (100) and the mutual distance d_{100}, two planes (010) with mutual distance d_{010} and two planes (001) with mutual distance d_{001}.

It is useful from reasons discussed below, to define instead of the "direct" lattice **a**, **b**, **c** a reciprocal lattice **a***, **b***, **c*** with the following specification:

$$\vec{a}^* = \frac{\vec{b}\times\vec{c}}{V_{EZ}} = \frac{\vec{b}\times\vec{c}}{\vec{a}(\vec{b}\times\vec{c})} \qquad \vec{b}^* = \frac{\vec{c}\times\vec{a}}{V_{EZ}} = \frac{\vec{c}\times\vec{a}}{\vec{a}(\vec{b}\times\vec{c})}$$

$$c^* = \frac{\vec{a}\times\vec{b}}{V_{EZ}} = \frac{\vec{a}\times\vec{b}}{\vec{a}(\vec{b}\times\vec{c})}$$

and the absolute lengths:

$$|\mathbf{a}^*| = 1/d_{100}; \quad |\mathbf{b}^*| = 1/d_{010}; \quad |\mathbf{c}^*| = 1/d_{001} \tag{3.99}$$

This means that the reciprocal lattice is defined as the cross product of two real-space vectors normalized on the volume of the unit cell, i.e. a reciprocal vector is perpendicular to two real space vectors and has the area between these vectors as absolute value. In Fig. 3.29 we may see the tripod of the reciprocal vectors constructed from the planes (100), (010) and (001) by dropping perpendicular to them. Hence, a translational vector in the reciprocal space has the following form:

$$\mathbf{t}^* = h\mathbf{a}^* + k\mathbf{b}^* + l\mathbf{c}^* \tag{3.100}$$

So, these hkl-values describe the crystal plane in the real space and lattice points in the reciprocal space. The sense of this—at first sight strangely seeming—construction is quite fundamental:

By reflection/interference of X-rays at lattice planes *hkl*, these planes transform to reflections on a screen/film fulfilling Bragg's law and can be indexed again with *hkl*'s as points of the connected reciprocal lattice. Hence, it is possible to construct geometrically the reciprocal lattice and to predict the location of the occurring Bragg reflections.

A very convincing representation of these properties has been given by P. P. Ewald (Fig. 3.30, according to Borchardt-Ott (1995)).

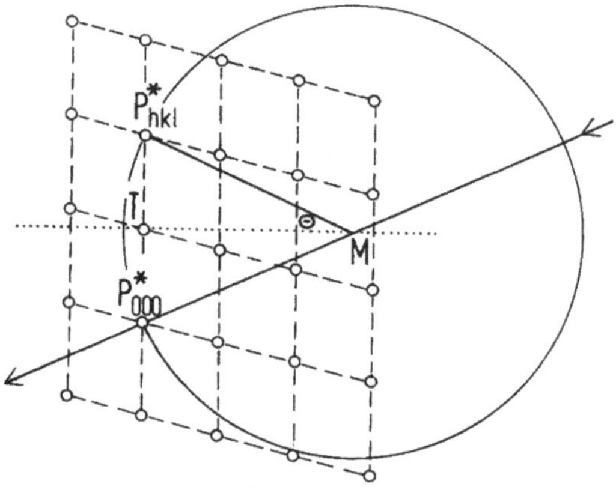

Figure 3.30 Ewald construction for a striking explanation of Bragg's law. The reciprocal lattice is turned together with the one in real space thus subsequently hitting the Ewald sphere with its reciprocal lattice points and producing a reflection. For a more detailed explanation of the geometry behind please see the text.

Imagine a two-dimensional reciprocal lattice with its origin in the point P^*_{000} (Fig. 3.30) and draw a line from this point in the direction of the supposed primary X-ray beam. Now let us assume a distance C/λ (C is a constant representing the scale of the drawing) and put this distance from P^*_{000} to the line of the primary beam to arrive at point M. From this point draw a circle around M with radius C/λ, which cuts the origin of the reciprocal lattice in P^*_{000}. Normally, P^*_{000} is the only point on the circle or surface of the sphere in 3D, respectively. The latter is called the reflection sphere. By a favoured position of the reciprocal lattice with respect to the primary beam it may occur that we will get another point P^*_{hkl} which is common of rec. lattice and reflection circle resp. sphere. In this case we would state that the Bragg condition is fulfilled with the direction of the scattered beam in M and P^*_{hkl}, since the triangle P^*_{hkl}, T, M (Fig. 3.30) can be constructed by

$$\sin\varphi = \frac{(1/2d)}{1/\lambda} = \lambda/2d. \tag{3.101}$$

From these geometrical properties we may conclude that each point of the reciprocal lattice within a sphere of radius $r = 2/\lambda$ may be brought into the position of reflection—this new sphere is called the limiting sphere. By this, we can calculate the number of possible reflections with the equation:

$$N = \frac{4}{3}\pi(2/\lambda)^3 / V^* = 33.5V/\lambda^3, \tag{3.102}$$

where V is the volume of the unit cell and λ the applied wavelength.

Hence, with decreasing λ the number of reflections will increase dramatically since the limiting sphere will become very large! For example, suppose a rhombic unit cell with metrics 8 Å × 10 Å × 20 Å. In the case of copper-Kα-radiation with λ = 1.5418 Å, we will observe 14,600 reflections, with a molybdenum Kα-tube (λ = 0.7107 Å) we will get 149,000 reflections! Also, the effective resolution of the reflection pattern will increase from 0.71 Å (copper) to 0.33 Å (molybdenum). Therefore, in the case of anorganic crystals with lattice constants comparable to the above ones, it is advisable to take a Mo-tube, in the case of organic crystals with their very large unit cells, the Cu-tube should be preferred.

Now, if we assume the value of λ to be fixed and θ variable, we are dealing with the Bragg turning-crystal method (monochromatic primary beam, the crystal will be turned). We may then assume by inspection of Fig. 3.30 that the crystal will be turned perpendicularly to the primary beam and hence, the reciprocal lattice will be turned equally around P^*_{000}. Then, other points of the rec. lattice will hit the reflection sphere and come into reflection position (= obeying Bragg's law). This procedure concerns all oscillation methods.

However, if we fix θ and change the wavelength λ (this is equivalent to a polychromatic X-ray beam and a static crystal), some λ's will be scattered out which fulfil with given θ and d_{hkl}

the Bragg condition—this will lead to "multi-coloured" reflections. In terms of the Ewald construction we then may assume that the reflection sphere will contain—due to the different wavelengths involved—a thick outer shell of the reflection sphere (like an "orange").

Here, the Laue method is concerned. As already mentioned above, the scattered rays lie on the surface of cone mantles and the Laue reflections are located on the intersection of these cone mantles with the projection plane.

The experimental generation of these Laue patterns is rather easy, but the evaluation, i.e. indexing of the different reflections, is not. It remains to ask whether there is a possibility to get the reciprocal lattice undistorted, i.e. receiving the intersecting cone mantles (hyperbola) as straight lines. The answer is: Yes, there is. This possibility will be shown in detail in Section 3.3.2 (X-ray diffraction).

3.3.1.5 Historical background

The X-radiation is a type of electromagnetic waves in the range of $\lambda \approx 0.25$ nm to 1 pm and $\nu \approx 0.25$–60 EHz. Hence, they continue the region of visible light towards shorter wavelengths, but, in contrast with these, we have no organ to detect the "X-Strahlen", as they were called by their discoverer, Wilhelm Conrad Röntgen on November 8, 1895, at the University of Würzburg, Germany. Therefore, in German language countries the name for these waves remained "Röntgenstrahlen", whereas in the Anglo-Saxon regions the original notation "X-rays" for the description of "until now unknown" type of radiation (X) survived. It is justified to ask how they could have been discovered, inasmuch we have no organ for them.

Obviously, one took the X-radiation as a sort of light[8] and used the recently invented photographic plates to record images of things in the path of the radiation. It is reported that probably the first exposure was the image of Mrs. Röntgen's hand showing the internal bones and the marriage ring on December 22, 1895. Perhaps this finding has been the reason that at first medical applications of the new type of radiation were investigated in detail. As it was found later, these experiments were not at all

[8]Therefore even nowadays, some people speak of "Röntgenlicht", which is a false characterization.

harmless: It turned out that this radiation is of ionising nature and may lead to cell damage in the human body with severe consequences. I remember reports of my late father about a doctor well known to his parents, who used the X-rays for diagnostic purposes and lost some of his fingers through amputation due to the almost permanent usage of "the tube".

The physical properties of the latter, which was invented by Röntgen, will be described in Section 3.3.2. However, the benefits of the X-rays surpassed the disadvantages by far, because from then on, pathologic changes in the human body could be detected in an early stage (e.g. one of the most widespread, main lethal diseases at those days, tuberculosis)—this saved many people. Still the X-ray diagnosis kept its importance till our present days (mostly applied as CT—computer tomography).

After his discovery, Röntgen became professor at Munich and received the first Nobel Prize of Physics in 1901. Up to then, the X-rays were used as a medium to produce shadows, i.e. differently absorbing material like, e.g., tissue and bones led to images of different greyscale with an amazing resolution.

As mentioned above, the genius idea of considering X-rays as waves with the property of being scattered at atoms or ions of a crystal lattice with subsequent interference was realized by Max von Laue, Walter Friedrich and Paul Knipping in 1912. Before that discovery, there was a discussion with famous Peter Debye, whether it should be possible at all to observe the assumed regular arrangement of atoms or ions in a lattice. Debye argued that the temperature motion of the particles would be so severe that a coherent scattering of waves could not be detected. Von Laue replied that the temperature motion would not occur stochastically, but in oscillations around a "centre of gravity", the later atomic or ionic positions within the unit cell.

The experiment of 1912 confirmed von Laue's opinion remarkably—for this discovery and its explanation von Laue was awarded the Nobel Prize of Physics in 1914. It was not only the birth date of X-ray diffraction and the possibility to thus determine the structure of arbitrary crystals, it also proved the wave nature of X-radiation that had been doubtful before.

Inspired by a publication of Max von Laue, Sir William Henry Bragg and Sir William Lawrence Bragg extended the studies of Max von Laue by investigating different crystals from 1912–

1914. They found the famous Bragg law mentioned above and developed the X-ray "spectrograph" to perform these measurements at crystal structures routinely. Both of them received the Nobel Prize of Physics for their merits in research discovering crystal structures.

A very broad and promising group of minerals are the silicates—so the here discussed example of fayalite Fe_2SiO_4 (Chapter 5). Sir W. L. Bragg developed a detailed theory of their structural properties together with his students in the following years and founded the practical side of diffractometry.

3.3.1.6 The Nobel Prize winner Max von Laue

Max von Laue was born on October 9, 1879, in Pfaffendorf near Koblenz, Germany. His father Julius von Laue worked in the German military administration and, like many other employees of the Services, often had to change his place of living. Thus, Max von Laue spent his youth at very different locations like Brandenburg, Altona, Posen, Berlin and Strassburg, where he was introduced to natural sciences. In 1898, after school, he absolved his military service for one year and then went to the University of Strassburg to study mathematics, physics and chemistry. After a stay at Göttingen and Munich, he went to the University of Berlin in 1902 and worked under famous Max Planck. After a two-year return to Göttingen after his thesis in 1903 on interference phenomena at plane-parallel plates, he started to work again at Berlin in 1905 as an assistant of Max Planck entering the Institute of Theoretical Physics devoting himself to thermodynamical topics. After his university lecturing qualification, he dedicated his time to the Theory of Relativity of Albert Einstein. In 1909 he moved to Munich as a Privatdozent at the Institute of Theoretical Physics at Ludwig-Maximilian-University (LMU), where he wrote one of the first textbooks concerning Special and General Theory of Relativity. However, his main merit was the discovery of X-ray scattering at crystals, together with Friedrich and Knipping in 1912. In the same year he became Professor of Physics at Zurich, later on (in 1914) at Frankfurt on Main. After World War I, where he did studies on telecommunication, he went again to Berlin as a Professor of Physics (1919) and worked together with Einstein at the Institute of Physics at Berlin-Dahlem; later on, he was also engaged as a consultant of PTR (Physikalisch-Technische Reichsanstalt) where my late father was also employed.

Max von Laue always had had a vivid sense of justice and fair play—he defended Einstein against the representatives of "Deutsche Physik" and was prematurely retired in 1943. With the repeated bombing of Berlin by the Allied Forces, von Laue went together with his institute to Hechingen (Southern Germany), was captured there by Anglo-American troops and interned in England together with nine other German scientists. However, his non-engagement or even critical attitude towards the political leaders during 1933–1945 was soon confirmed and acknowledged; shortly after the end of War, he was allowed to go to Göttingen as Acting Director of the Max-Planck-Institute and Titular Professor at the University there.

In 1951 he got an appointment as the Director of the Fritz-Haber-Institute for Physical Chemistry at Berlin-Dahlem, where he was mainly working on X-ray optics till his retirement in 1958.

Contrary to many of his colleagues, his scientific knowledge was amazingly broad and, at the same time, profound. As mentioned above, he started with thermodynamics, then Special and General Theory of Relativity, X-ray diffraction, Theory of Superconductivity, History of Physics a.s.o., always not remaining "at the surface", but providing excellent publications on the actual subject, even monographs. His honours and honorary doctorates are numerous and would need an own chapter to describe, and, at the same time, his interests were not confined to physics and natural sciences. He was fond of sailing, mountain hiking, skiing and motor sports. The author of this book, perhaps by some strange "connection of souls" or just by chance, is also practising these hobbies, even the one of hurrying to his lectures on a motorbike and driving a sports car. As an irony of fate, Max von Laue, who never had had any accident before, was hit in his car on the AVUS, a famous speedway at Berlin, by a completely inexperienced motorcyclist and died, after having seemed to recover, on April 24, 1960. It is well known that von Laue also loved the fine arts, in particular classical music. Very often he visited my late father in his Laboratory of Musical Acoustics at Physikalisch-Technische Bundesanstalt and urged him to play "etwas Schönes" on his Hammond organ or on his

HiFi- device. Then he was sitting in a big armchair and enjoying the music. This chair now enlightens my office and I urge only dear friends or my most esteemed colleagues to take a seat there.

3.3.1.7 Later contributions

As mentioned above, several methods of diffractometry have been developed after Max von Laue's discovery in the course of time. Already in 1913, the method of Sirs W. H. Bragg and W. L. Bragg for the application of their famous Bragg equation has been established:

$$2d \text{ sine } \varphi = n\lambda \tag{3.103}$$

By evaluation of this formula, we might easily distinguish between the possible basic methods of diffractometry: The oldest one stems from Max von Laue; the wavelength λ is "allowed" to vary by taking the "white" primary X-radiation without filtering or monochromatization. Therefore the involved crystal has to be kept fixed (φ is constant) and the reflections selected according to the scattered wavelengths. It is easily possible to derive symmetries, but hardly d-values or lattice constants.

Bragg's method keeps the wavelength λ fix (monochromatic), the angle $\varphi = \theta/2$ is varied. It is possible to derive different d-values; it is an oscillation method and allows to determine distances of net planes d and lattice constants.

Almost at the same time as Father and Son Bragg (1913), Maurice de Broglie used an oscillation method to examine the radiation of an X-ray tube. Later on, Hugo Seemann (1919), Michael Polanyi, Ernst Schiebold and Karl Weissenberg (1920s) made further developments.

The principle of the oscillation method is based on the premise that a lattice axis in a single crystal be aligned parallel to the vertical axis of a hollow cylinder, whose wall is covered by a film. For the measurement, a monochromatic X-ray beam is laid perpendicular to this cylinder axis and the crystal turned around the latter by a motor.

For the understanding of the resulting image on the film, it is best to call the construction of the Ewald sphere in mind (cf. Section 3.3.1.4). Let us assume to have the c-axis of the crystal being vertically aligned: By turning the crystal we will obtain concentric Laue cones which will intersect with the cylinder wall. On the deconvoluted film there are parallel reflection lines visible whose distances will give, by calculation, the lattice constant c_0.

However, it is not possible by this method to derive an indexing of the reflections or to gain more detailed information about the metrics. In order to align the crystal with a lattice vector parallel to the cylinder axis, oscillation exposures have to be performed, where the crystal is only tilted by angles between 5 and 15°.

A further progress in oscillation methods was achieved by the Weissenberg procedure. Contrary to the basic method described above, a cylindrical layer diaphragm is inserted in front of the cylinder wall with the film, by which the reflections of a special Laue cone can be selected. Additionally, the motion of the crystal is coupled with a translation of the film, so that the reflections on the lattice lines will be expanded and can be indexed much more easily (Weissenberg 1924). We may understand this method as the ancestor of the BPC method (see Section 3.3.2) and it was rather widespread among crystallographers. A colleague of the author even preferred this old procedure, since it is able to record several, e.g. *hk*1, *hk*2... planes at the same time. A disadvantage is the strong distortion of the lattice lines leading to "garlands" which have to be evaluated with special drawing tools.

Already in 1916, the Debye–Scherrer method was developed. It was invented by Peter Debye, Paul Scherrer and, independently, by Albert W. Hull. Calling again the Bragg equation in mind, we again need a monochromatic X-radiation, but instead of a single crystal, it is equivalent to take a crystalline powder sample. The mechanism of the device was already explained above, we confine ourselves to the statement that the statistical arrangement of the crystallites will always lead to a "favoured" orientation of some of them (in the sense of the obedience of Bragg's law)—the calculation of the glancing angle can be performed from the location of the Laue cone intersections on the film and, subsequently, the calculation of *d*-values established. The latter are listed for each measured compound, in the old days of powder diffraction on a large compilation of cards, nowadays the procedure is automated by help of a data bank (ICSD). The main application is the detection of phases from mixtures of phases from an unknown powder sample. For this, "combs" of intensities and related d-values of assumed single compounds are taken from the data bank and laid over the measured total diffractogram. On a computer screen, the success or failure of the assumed "comb" can be instantly followed and the suspected contribution either confirmed or ruled out. By the superposition of several of these "combs" a complete phase determination of some unknown powder sample will be possible.

With the advent of powerful computers, even the profiles of the Debye–Scherrer peaks could be refined (Rietveld analysis). Nowadays, even structure solutions and very exact determination of lattice constants have become possible.

Another method of the same kind but with another geometry is the procedure after Guinier. The cylindrical sample and film chamber with the powder sample in the centre of the device is replaced by an off-centre construction: The sample is placed at the inner wall of the chamber, like the film itself. Each ray traversing the chamber and being focused on a point on the wall will be scattered at the other wall and focused at the same point. This geometric construction is similar to a spherical mirror at which an incident beam does not need to be parallel but may be focused by this mirror through grazing incidence (X-rays do not obey the normal law of refraction but have a refractive index for all materials of 1). By this measure, the intensity of an X-ray source can be used much more economically—one can take exposures in transmission and back-scattering. Moreover, the reflection lines are more narrow than those of the Debye–Scherrer Camera. As mentioned above, the film may be replaced by a moving counter tube, always obeying a φ-variation for the sample and a 2θ one for the lever of the counter, so that the Debye–Scherrer rings are intersected at one (diffuse) point by the aperture of the counter tube thus obeying Bragg's law.

Later on, the method has been automated up to the stage that nowadays, by a sort of "elevator", different samples can be measured one after the another, so that the powder patterns of whole series of compounds can be recorded continuously.

Turning again towards the single crystal methods, the invention of the BPC (Buerger Precession Camera) by Buerger (1941) was a considerable progress. A BPC uses a monochromatic beam and allows to determine lattice parameters (metrics) and even the space group of the given crystal by evaluation of the different layers (net planes) and systematic absences of reflections. Since the deconvolution of the former cone mantles and their intersections leads to linear lattice lines just by mechanical means, the obtained image of the reciprocal lattice is undistorted and can be used, as in our case, as a convenient tool for crystal orientation (cf. Section 3.3.2.1).

However, also the theory of diffraction kept path with the experimental progress.

One of the most important highlights in modern physics of the 20th century was the discovery of Louis-Victor de Broglie (the younger brother of the Maurice de B. mentioned above) about the wave nature of classical particles—in particular the one of the electrons.

In his famous thesis about quantum theory he drew the unusually courageous conclusion that the wave—particle dualism should be applied to any form of solid matter. This fully agrees to the theoretical work of Erwin Schrödinger who defined "matter waves" for his famous equation (cf. 3.2.1.1), almost at the same time period. Up to then, the conclusion of de Broglie was not confirmed experimentally, and some of his elder colleagues were sceptical about this uncommon idea. However, experiments of Davisson and Germer 1927 and of G. Thomson in 1928 showed interferences of electrons at scattering structures and confirmed remarkably the postulate of de Broglie, for which he was awarded the Nobel Prize of Physics in 1929. In detail, de Broglie attributed an infinite wavelength to a fixed particle; application of the Lorentz transformation to a particle in motion leads to a wavelength $\lambda = \hbar/p$ with p being the momentum (impulse) mv of the particle (a classical particle quantity), whereas λ describes a typical wave property. Hence, both properties are connected by Planck's constant, so that an old physical problem had been elucidated.

Concerning our topic "diffraction" this meant that with any of the former classical particles, diffraction experiments should be possible, including neutrons.

These particles have been detected earlier, but not recognized as such. The identification of the neutron as a constituent of the nucleus was provided by James Chadwick, who received the Nobel Prize of Physics for this discovery in 1935.

The reason for mentioning all these features here is the fact that neutrons, besides their importance for nuclear physics, are eminently suitable for diffraction purposes. This is due to the observation that the de Broglie wavelength of thermal (i.e. slow) neutrons lies in the order of the diameter of atoms/ions and distances between them. Hence, it is possible to perform neutron diffractometry—the details will be presented in a special chapter below (Section 3.3.4). This great possibility to investigate structure and physical properties of matter was mainly realized by Clifford Shull and Bertram Brockhouse in the 1950s. For this pioneering work they were awarded the Nobel Prize of Physics in 1994.

Since neutrons are sensitive to the nuclei (contrary to X-rays which are scattered by electrons) and to unpaired "magnetic" electrons, an overwhelming number of crystallographic and magnetic structures have been solved with neutron scattering.

However, again, the theoretical findings kept pace with the achievements of experimental diffractometry. In Section 3.3.2.3 the phase problem will be mentioned, a severe obstacle to determine crystallographic structures with the recording of intensities without the possibility to derive the phases of the involved structure factors as well.

From 1953 to 1956 Herbert Hauptman and Jerome Karle developed a statistical method for the direct determination of crystal structures by analysis of X-ray scattering, the so-called "direct methods". For these findings they received the Nobel Prize for Chemistry in 1985. The direct methods are based on the premise that the amplitude and the phase of a wave are not in any case independent on each other but that in the case of structure factors both may have certain connections between each other—the phases can then be derived directly from the structure amplitudes. Therefore this method, which is based on statistical considerations, is called "direct method" (valid for up to 50–100 atoms per cell).

Another possibility to solve the phase problem is the Patterson analysis, mainly applicable in the presence of heavy atoms.

However, we will not go into further detail here, since the structure solution was of minor importance for our nanoscope: We are mainly interested in the collection of structure factors and phases of already solved crystallographic structures.

With the vast progress in computer technology starting from the 1960s, it has become possible to solve structures just by "trial and error". There, calculated structure factors F_{cal} will be derived from probable atomic coordinates from a predefined model by using the equation

$$F_{cal} = F_{hkl} = \sum_{j}^{N} f_j e^{2\pi i(hx_j + ky_j + lz_j)}. \tag{3.104}$$

Then, these will be compared with experimental structure factors F_{obs}. By varying the atomic coordinates ("refinement") in an iterative procedure until the difference between F_{cal} and F_{obs} will drop below a predefined minimum, we claim that "the structure is refined".

In particular, starting from the 1980s, where sufficiently powerful workstations or even personal computers entered modern crystallography, this mode of refinement was "state of the art".

Nowadays, diffractometers normally are equipped with an extra PC and a suitable software (e.g. SHELX, one of the most popular programs by Sheldrick (1997)) to solve structures from the incoming intensity data. An immense experimental progress besides neutron scattering was established by the increasing usage of synchrotron radiation for diffraction purposes. Again, I will get into more detail in a special section (3.3.3). Here, it may be sufficient to claim that with the operation of large particle/electron accelerators (e.g. DESY at Hamburg, Germany) crystallographers dispose of a very intense X-ray source of extreme goodness (narrow width of the involved frequency line) which can be used for diffraction purposes. In particular, for tiny crystals very distinct electron density maps, phase transitions, multipole refinements a.s.o., synchrotron diffractometry is the method of choice. We also profited from this comparably new tool for the evaluation of outstanding structure factors and phases for our nanoscopic evaluation of synthetic fayalite.

3.3.2 X-Ray Diffraction

3.3.2.1 The Buerger Precession Camera

Although many types of X-ray cameras for diffraction purposes have been invented in the course of time, the most "popular" one certainly is the Precession Camera invented by Buerger (1940).

The great advantage is that the reciprocal lattice is displayed undistortedly with more or less "directly" measurable distances and angles in order to obtain the metrics of the sample's unit cell. Since we used this technique to orient and to cut our fayalite crystal sections for Mössbauer spectroscopy, some more space is dedicated to this method here. In order to obtain a rather complete impression of the procedure, let me describe the whole Buerger Precession Camera (BPC) with all necessary peripheral parts.

The main condition to operate a BPC is to have a sufficiently powerful X-ray tube. For this purpose, we need a high-voltage generator which may provide a sufficiently high tension (in the range of up to ≈50 kV) and a current of several mA, so that a common X-ray tube produces a power of several kW. In the times of von Laue, Bragg, Buerger and other famous crystallographers till our present days these tubes have been water-cooled; however, nowadays it is possible to construct air-cooled tubes or very effective ones with a comparably low power. Such a device is shown in Fig. 3.31—it has the great advantage that all important parts are visible here.

Figure 3.31 Air-cooled X-ray Mo-tube from the XR 4.0 expert system of PHYWE Ltd., Göttingen, Germany. Through the evacuated glass cylinders without water cooling or other outer screens the bright spot, where the electron beam hits the molybdenum target, can be nicely seen from the outside.

In an evacuated glass tube, electrons are generated in a heated filament and accelerated in a strong electric field versus a metallic target. In the target whose material is decisive for the wavelength of the X-rays, the electrons are "braked down" by multiple collisions with the atoms of this target (anode). Due to the strong intensity of the internal electron beam, the anode has to be cooled internally in order not to melt. By the bombardment of the target two sorts of X-rays are released perpendicular to the tube axis:

(1) A "white" radiation of continuous wavelengths with a calculable minimum $\lambda_{min} = 12.398/V_{acc}$ and a maximum at $\lambda > \lambda_{min}$ (see Fig. 3.32, left).
(2) A characteristic radiation due to electronic transitions inside of the target atoms, depending on the target material (see Fig. 3.32, right). With increasing order number of the target elements we obtain increasingly shorter wavelengths.

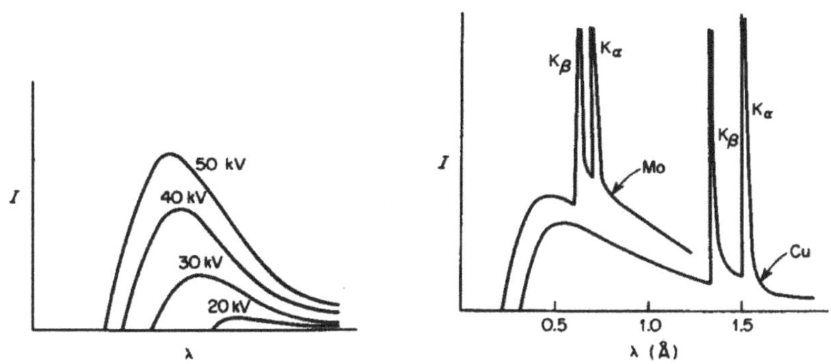

Figure 3.32 Dependence of the X-ray intensity dependent on the wavelength with different acceleration voltages as a parameter. The intensity distribution is called "bremsstrahlung" (left image). On the right hand side the so-called "characteristic radiation"—to be seen as spikes on the bremsstrahlungs-background—is shown for different target materials (Mo and Cu).

The transitions stem from different atomic shells and, for purposes of diffractometry, it is sometimes troublesome to obtain multiple different wavelengths (K_α, K_β, ...). In order to obtain a rather distinct wavelength, we have to attenuate or suppress the other ones.

Two principal methods are commonly used for this purpose:

(1) The metallic foils of several elements display an exponential dependence on X-ray absorption with strong discontinuities dependent on the wavelength, the so-called absorption edges. If we superimpose such an edge of a suitable material foil on a non-desired peak of characteristic radiation, it is possible to suppress the latter almost completely.

(2) As already mentioned above, a very fine crystal of a suitable material with minimum failures can be set to a special angle fulfilling Bragg's law for a distinct high-intensity reflection. The connected wavelength from this *hkl* plane/reflection can be used as the new "primary beam" with a nicely sharp wavelength (monochromator). The high goodness, however, is won against a loss of primary intensity.

Figure 3.33 shows a conventional BPC from Huber Ltd., Rimsting, Germany. Generally speaking, it is based on an osc. sample/oscillating film procedure. In order to understand its mode of operation it will be best to explain the essential parts of the BPC using Fig. 3.33.

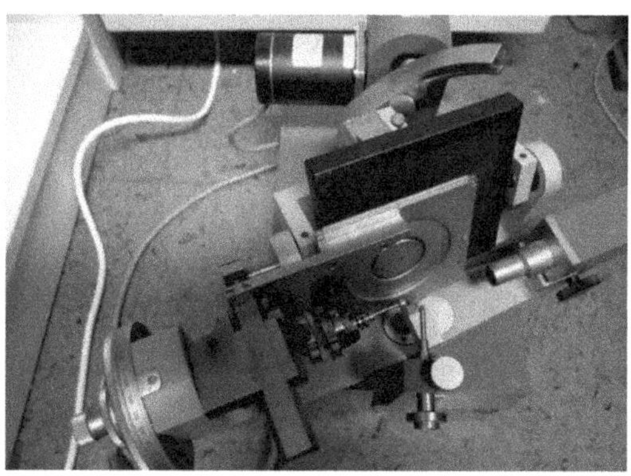

Figure 3.33 Photograph of a Buerger precession camera (BPC) from Huber Ltd., Rimsting, Germany. The single parts of the camera are clearly visible and explained in the text.

The BPC is shown in the figure without the necessary X-ray tube—so we may notice at the bottom of the image the "open" collimator, from where the X-ray primary beam will hit the single crystal. The latter is mounted on a so-called goniometer head which is shown separately on Fig. 3.34, left.

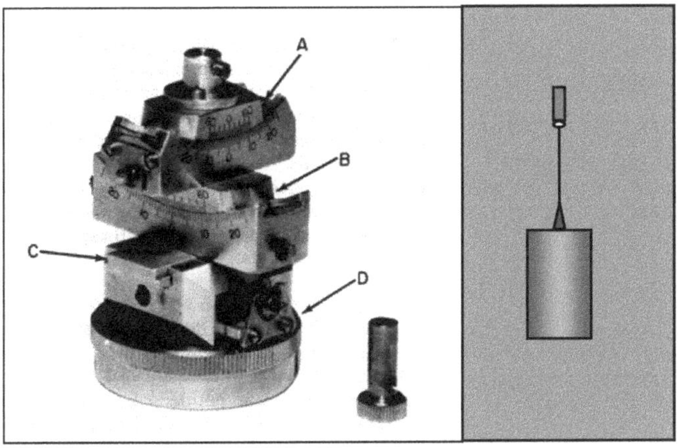

Figure 3.34 Standard goniometer head with the arc sledges and angle grading together with nonius teilung (A, B) as well as translational sledges (C, D) to compensate for decentring of the crystal during arc adjustment. The image on the right hand side shows the crystal bearing unit (to be mounted at the top of the goniometer head) with brass cylinder, inserted glass fibre, glue and crystal sliver.

For the mounting of small crystals, usually a piece of a glass capillary is glued into the central hole of a small brass cylinder and the crystal subsequently attached to the other end of the capillary by a tiny drop of a low-viscosity glue (i.e. cyanoacrylate gel) under the microscope (Fig. 3.34, right). For bigger crystals, say, those with dimensions of $5 \times 5 \times 1$ mm^3 and larger, a paraffin wax—which becomes less viscose by hand warming—is to be preferred, since the crystal can subsequently be easily removed from the wax. These arrangements are mounted at the top of the goniometer head. The arc sledges with angle tics directly below the crystal are intended for the lateral turning of the crystal, the arc following below is for a left-right tilting with respect to the beam. The other (straight) sledges without angle tics are intended for a translational correction of the crystal, since it may happen that, by large angles, the crystal will leave the beam. The carabiner lock at the bottom of the goniometer head serves to attach the whole device unambiguously to the spindle of the big wheel of the BPC with angle tics for up-down tilting of the crystal. It may be seen at the left side of Fig. 3.33. On the right side we have a small telescope with a crosshair; turning the big wheel should always leave the crystal with its top (the crosshair indicates the location of the beam) at the same point in space, for—during motion of the BPC—the crystal must not leave the beam. Small samples should "bath" in the X-rays, larger crystals should be mounted in "grazing incidence", since sometimes the bulk of the sample will severely absorb the radiation. Versus the top of Fig. 3.33, we see the diaphragm holder, where necessary big plates of ring diaphragms (Fig. 3.35) can be inserted.

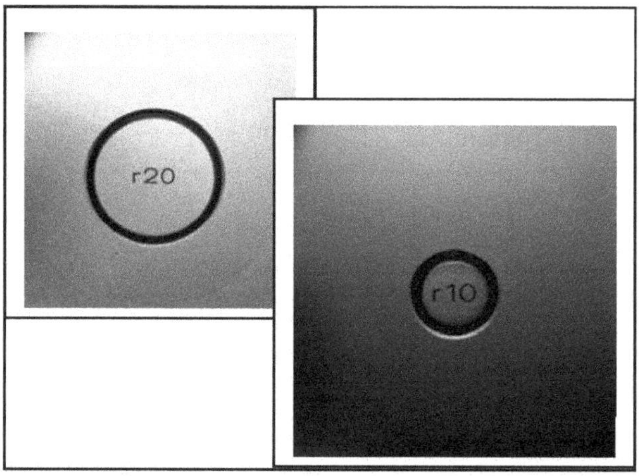

Figure 3.35 Two examples of metallic diaphragm plates omitting a ring with radius of 20 and 10 mm, respectively.

This holder and also the film cassette behind the former one are connected through a lever system and ball-and-socket joints with the big wheel. It is possible, at the horizontal part of the lever system, to change the distance between the crystal, the diaphragm holder and the film cassette by a protractor. The film cassette also can be shifted by a small mm-scale that may be noticed in Fig. 3.33 above the dark cassette. Above the whole assembly we see a big arc sledge, where the whole device (crystal, diaphragm and film), can be tilted by a variable angle μ. This arc is attached to a motor, so that the device will perform a strange, precession-like motion—therefore the name "Precession Camera".

The geometric properties, however, are best illustrated by drawings. Let us assume that a lattice vector [100], [010] or [001] be parallel to the primary beam (Fig. 3.36, top left).

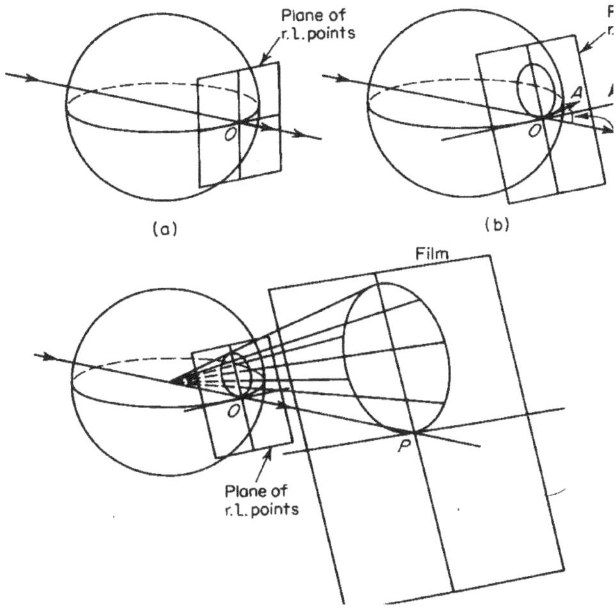

Figure 3.36 Top left: Lattice vector parallel to the primary beam in the origin of a reciprocal lattice plane hitting the Ewald sphere. Top right: Inclination of a reciprocal lattice plane by an angle μ with respect to the primary beam. Bottom: Film cassette inclined by the same angle μ. Crystal, film cassette and diaphragm (see below) perform a complicated precession motion around the initial reciprocal lattice vector.

In this position we will observe no reflections on the film. When the crystal now is tilted by an angle μ around the axis A perpendicularly to the primary beam, the reciprocal lattice plane (say, $(hk0)$ in the case of lattice vector [001]) will intersect Ewald's reflection sphere in a circle (Fig. 3.36, top right). By turning the crystal in a mode that AO will make a precession around the primary beam by an angle μ, this intersection circle will also precede, hit the reflection sphere and will permanently provide reflections by fulfilling Bragg's law—one may call this motion "scanning with a circle". The scattered secondary rays lie on a cone mantle whose intersection with the reflection sphere produces the circle.

The question is how this preceding circle can be deconvoluted. It is done by moving the film with the same precession motion as the crystal (using the big arc).

There is a nice, convincing comparison for this procedure: Imagine a helicopter with a movie team filming a car on a curvy street from close distance; viewing the movie later, it will not be possible to perceive the curves—the car is meant to run linearly. Here, the former circle will be deconvoluted into single lattice lines.

There is one problem: We have considered the "zero-layer" (*hk*0), but other planes will provide contributions either, since the reciprocal lattice is not flat consisting only of (*hk*0), but extended like a "thick board". It is necessary to suppress the other reciprocal planes so that only (*hk*0) will hit the film. For this purpose we will need the ring diaphragm. There are several sizes available (dependent on the distance between film and reflection sphere, i.e. origin O – P^*_{000} of the sphere). The geometrical properties are shown in Fig. 3.37.

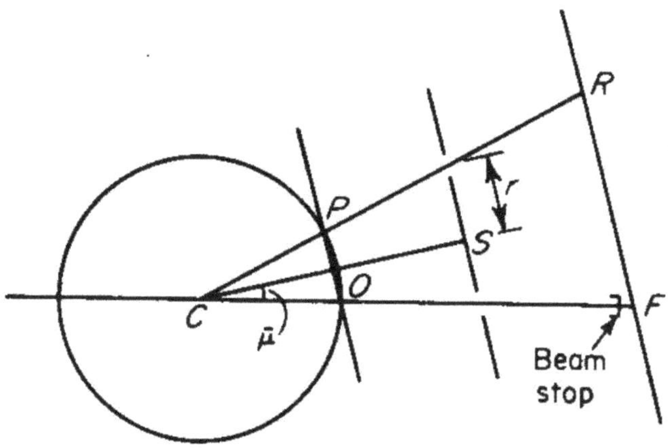

Figure 3.37 Buerger precession geometry, side view. CF is the crystal-film distance, RF the distance from the centre of the film to its boundary and *r* the radius of the diaphragm.

The ring diaphragm then must precede like the film in order to fade out exactly one cone (belonging to the plane (*hk*0)). From Fig. 3.37 it is evident that we need other diameters of the ring(s) and other distances of the film for selecting other planes *hk*1, *hk*2, a.s.o.

Generally speaking, we may distinguish between three principal parameters:

(1) The distance crystal-film CF ("focal distance"). At most commercial BPC's this distance CF amounts to 60 mm. This can be verified easily:

The distance from the centre of the film to its boundary amounts to 60 mm; according to the law of rays it is evident that

$$RF/CF = PO/CO = 60/60 = 1 \quad (3.105)$$

This means that PO is the radius of the reflection sphere. By setting the radius to the circumference of the circle one obtains an arc of 60° in the centre and for the other angles of the triangle.

(2) Half of the above 60°-value is 30°. This is the maximum possible value for μ. Smaller angles are possible, but they correlate with other distances CS_0 and radii of the diaphragm ring. It is evident that a ring with comparably small r (and μ) close to the crystal will give the same cone like a ring with larger r (and μ) more distant from the crystal (Fig. 3.37).

Therefore, the following connection is valid:

$$r_B/CS_0 = \tan\mu; \; CS_0 = r_B/\tan\mu, \quad (3.106)$$

where r_B is the diaphragm radius, μ the precession angle and CS_0 the distance of the diaphragm with respect to the crystal. This is also the translation which has to be adjusted at the lever between the diaphragm holder and the big wheel axis (cf. Fig. 3.33). After choosing the relevant values for r_B and μ it must always be tested, whether parts of the BPC may collide with the tube housing. In this case other values for μ and r_B have to be selected!

Example: With μ = 30° and r_B = 20 mm we will get a diaphragm distance of 34.6 mm, which will not lead to any collisions.

Hence, with this adjustment it is now possible to record a virtual zero layer (*hk*0-plane). For that, one has to switch on the motor drive and let expose the film during around one day. At the tube housing a suitable filter has to be selected to produce a monochromatic beam.

In the ideal case, we will obtain on the developed film a linear pattern of reciprocal lattice points which correspond to the connected lattice planes in the real space. It is remarkable that through the precession, cone-like motion of crystal, diaphragm and film the normally occurring hyperbolic lines are dconvoluted to linear lattice lines with [*h*00] and [0*k*0] on horizontal and vertical axes on the film, respectively. Within the four quadrants we observe *hk*0 reflections of the "zero-layer" mentioned above. We now may calculate a reciprocal distance d^* [Å^{-1}] by measuring one of the occurring distances d_{Film}, e.g. using a ruler. The very easy formula is

$$d^* [\text{Å}^{-1}] = d_{\mathrm{Film}}\,[\mathrm{mm}]/\,(\lambda[\text{Å}] * M[\mathrm{mm}]), \tag{3.107}$$

where λ is the applied X-ray wavelength and M = focal distance of the camera (= 60 mm).

Let us consider our main example of synthetic fayalite; in this case, however, we recorded the plane 0*kl* as zero layer:

With molybdenum radiation (λ = 0.7107 Å) the following distances on the film have been measured (Fig. 3.38):

$d_{1,\mathrm{film}}$ = 8.75 mm and $d_{2,\mathrm{film}}$ = 6.75 mm.

From the above formula we get

$$d_1^* = 8.75/0.7107{*}60 = 0.205\ \text{Å}^{-1}$$
$$d_2^* = 6.75/0.7107{*}60 = 0.158\ \text{Å}^{-1}. \tag{3.108}$$

Since the crystal was rhombic, we only have to take the inverse of the previous values in order to obtain two lattice constants in the real space:

$$d_1 = 1/|0.205| = 4.87\ \text{Å} \quad d_2 = 1/|0.158| = 6.82\ \text{Å} \tag{3.109}$$

These values correspond well with the lattice constants b_0 and c_0, respectively, of synthetic fayalite (space group Pnma).

Taking our initial example of the *hk*0 layer, we would have to translate the position of the film by one reciprocal lattice constant c_0 in order to record a *hk1* layer. For that, we ought to know this constant c_0 and to transform it to its inverse value by

$$V_1 = \lambda M/d_0, \tag{3.110}$$

where d_0 is c_0 that can be obtained by another "zero layer", e.g. the one above (fayalite). However, by this translation of the film cassette, we also will get a modified precession angle Ψ_1 and another diaphragm distance *s*. This would lead too far, but it may be kept in mind that all lattice constants and angles can be determined by subsequent precession exposures.

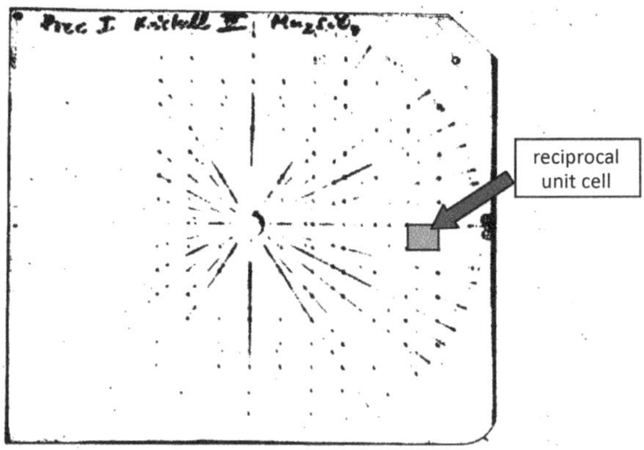

Figure 3.38 Negative of a film sheet of a fayalite Buerger precession exposure of the zero layer *hk*0. The grey-coloured rectangle denotes a projection of the reciprocal unit cell into the (*a*, *b*) plane giving the rec. lattice distances |**a***| and |**b***|.

Mostly, the crystal has to be fixed perpendicularly to the initial position. Another issue is that we do not necessarily have a lattice vector parallel to the beam as mentioned in the beginning of our consideration. On the contrary, this would be pure luck. Fortunately, however, it is possible to get such a direction on the same device by so-called orientational exposures.

For this purpose, "little" precession exposures with a minimum precession angle μ of 9.5° are taken with an unfiltered ("white") beam. We will then get after almost half an hour exposures with a strange view: A small circle is visible with reflection streaks of different lengths (Fig. 3.39, left):

It looks like seeing towards the bright end of a dark tube. In most cases, the inner "circle" is off-centred. It would lead too far to discuss how to calculate the procedure of centring—in fact this is performed by turning the different goniometer head arcs and the big wheel of the BPC—but the aim of the procedure is to get this inner circle centred. When this will happen, we can be sure to have a lattice vector parallel to the primary beam. From then on we may start the whole BP procedure as mentioned above.

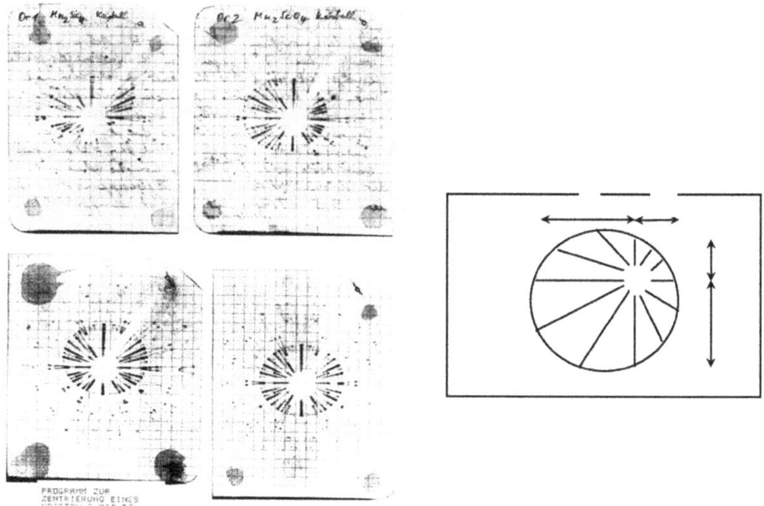

Figure 3.39 Left image: Negatives of orientation images with a precession angle of only 9.5°. We might consider them as tiny precession exposures. Right image: Schematic view of the reflection streaks with the crystal being misoriented. By observing the streak differences along the double arrows it is possible to transform these differences into the necessary corrections of the goniometer head arcs and the handwheel of the BPC in order to get the exposure successively centred and to have a lattice vector parallel to the X-ray beam. This is the prerequisite to start a BPC zero-layer exposure (Fig. 3.38).

Another obstacle is often observed as "empty spots" on the precession exposure *hk*0 where there should be a reflection. Certain symmetry elements in the crystal lattice like glide planes, screw axes and centrings lead to extinction of reflections by destructive interference. The pattern of these regular ("systematic") extinctions as a function of *hkl*, however, is characteristic for the space group involved. From these patterns in different directions of space we even can deduce this space group of the considered crystal. Having evaluated the latter and metrics, however, we may even turn towards the intensity of the different *hkl* reflections—in the early days of crystallography this has been done photographically by following the gradual weakening of the reflection intensities by a grey wedge. In the more recent days, however, say, around the 1960s, more sophisticated methods have been applied.

3.3.2.2 The intensity distribution of the reflections

For a full determination ("structure solution") of a given compound we also need the intensities of the different reflections. As pointed out above, it is possible by such investigations to obtain the assembly of the unit cell itself i.e. the positions and type of atoms or ions within the cell. This can again be derived from the effects of interference, but on a finer level than before.

Let us consider a CsCl lattice (body-centred cubic, bcc) consisting of a primitive lattice of Cs^+ ions and another primitive lattice of Cl^- ions, translated by half of the space diagonal against each other (Fig. 3.40).

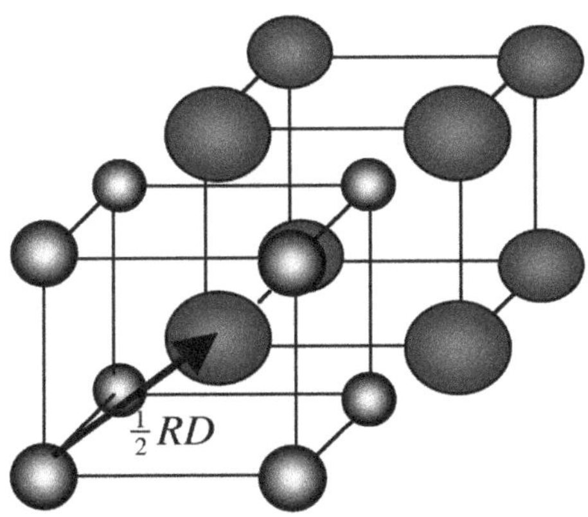

Figure 3.40 CsCl lattice, schematically, with small Cs^+ ions (bright metallic) and Cl^- ions (dark grey).

Two X-waves arising from each of the two sublattices would produce interference on the condition that the path difference would amount to

$$\Delta = 1/2\,(h_1 + h_2 + h_3)\,\lambda \qquad (3.111)$$

since we get by translation of $t_1 = t_2 = t_3 = 1/2$ from one sublattice to the other.

Now it depends on the special value of the triple $h_1h_2h_3$ (or *hkl*, respectively), whether the superimposed intensities are added or subtracted. By the general rule of interference we obtain:

For $h_1 + h_2 + h_3$ = even (amplification) and
for $h_1 + h_2 + h_3$ = odd (weakening) (3.112)

of the reflections involved. Because of unequal electron numbers of Cs^+ and Cl^-, respectively, we do not observe complete extinction. Hence, we record strong "even" (e.g. 220) and weak "odd" (e.g. 210, 230) reflections.

Instead of the addition of partial waves from different sublattices, we just have to add the partial waves of the atoms to construct a cell wave and, by summarizing over the cell waves, we will get the lattice wave.

A very important quantity here is the so-called structure factor. It is defined as the scattering amplitude of the atoms involved with respect to the scattering amplitude of one electron, namely for our above example:

$$F_{h1h2h3} = \frac{A_e}{A_{e^-}} = \frac{A_{Cs^+} + A_{Cl^-}}{A_{e^-}} = f_{Cs^+} + f_{Cl^-} \quad \cdots \quad \text{for } h_1 + h_2 + h_3 = \text{even} \qquad (3.113)$$

$$F_{h1h2h3}' = \frac{A_o}{A_{e^-}} = \frac{A_{Cs^+} - A_{Cl^-}}{A_{e^-}} = f_{Cs^+} - f_{Cl^-} \quad \cdots \quad \text{for } h_1 + h_2 + h_3 = \text{odd} \qquad (3.114)$$

where f is the atomic form factor.

The latter quantity describes the slope of the scattering ability of a specified atom or ion. It is correlated with the decrease of the radial part of the wavefunction due to the electron distribution around the given atom/ion. This form factor has been calculated for many elements by Watson and Freeman (1961) and is approximated by a power development of the connected function.

3.3.2.3 Generalization of the above example

Let us assume to have s different atoms per unit cell with the local translation vector $\mathbf{r}_k$, k denoting the k-th sort of atoms:

$$\mathbf{r}_k = \rho_k \mathbf{a} + \sigma_k \mathbf{b} + \tau_k \mathbf{c} \text{ with } \rho, \sigma, \tau < 1 \qquad (3.115)$$

According to a common usage, ρ, σ, τ are the fractional coordinates of a given atom or ion in the unit cell and are expressed as fractions of the lattice vectors (as such they are dimensionless).

The path differences of the X-waves stemming from the atoms k may then be written as

$$\Delta_k = (\rho_k h_1 + \sigma_k h_2 + \tau_k h_3)\lambda \qquad (3.116)$$

The structure factor can then be expressed as a sum over the k different form factors in the cell times a phase factor stemming from the different atomic positions within the cell:

$$\mathrm{F}_{h1h2h3} = \sum_{k=1}^{s} f_k e^{i\varphi_k} = \sum_{k=1}^{s} f_k e^{2\pi i(\rho_k h_1 + \sigma_k h_2 + \tau_k h_3)} \tag{3.117}$$

where f_k is the form factor and $e^{i\varphi_k}$ the phase factor.

Whereas the structure factor is a somehow abstract quantity, the intensity of a reflection is not: It can be measured, e.g., by the local blackening of a film visible as a reflection spot and a characteristic distribution of the blackening (profile); the integrated intensity over the volume of the spot will give the total intensity. Another, more common procedure nowadays, is the total count rate of X-ray quants at the position of the reflection. In any case, the intensity of a reflection $h_1h_2h_3$ is proportional to the structure factor squared—since the latter is a complex quantity, the real structure factor part F has been multiplied by its conjugate complex part F^*:

$$I_{h1h2h3} \propto |F_{h1h2h3}|^2 = \mathrm{FF}^* \tag{3.118}$$

However, in centrosymmetric cases, we may take the real parts only and can replace the structure factors by simple sine and cosine functions:

$$I_{h1h2h3} \propto |F_{h1h2h3}|^2 = \left(\sum_{k=1}^{s} f_k \cos 2\pi(\text{Phase})\right)^2 + \left(\sum_{k=1}^{s} f_k \sin 2\pi(\text{Phase})\right)^2 \tag{3.119}$$

The proportional sign in the above equation outlines several quantities as, e.g., Lorentz-, polarization, absorption factor a.s.o.—it would lead too far to discuss the details here. As a general remark, it may be stated that by measurement of the intensities of the reflections *hkl* it is possible to determine the fractional coordinates ρ, σ, τ for each atom/ion within the unit cell.

Unfortunately, this relation is not unambiguous; it is impossible to unequivocally derive the phase (the term in round brackets in the above equation) from the intensity of the reflection. This issue has been widely discussed in literature and has been denoted as "the phase problem".

Several remedies have been proposed to treat this obstacle—it would again lead too far to enter into the details here; a good review of the topic "structure determination" is provided by the famous book of Stout and Jensen (1989).

Let me again address the form factor f_k. It has a general slope proportional to sineθ/λ, hence,

$$f_k = f(\sin\theta/\lambda) \tag{3.120}$$

so that we may consider the form factor as an electron density, since the scattering contributions of the different electrons look like a diffuse sphere or better: hyperarea depending on the three coordinates x, y, z (ρ, σ, τ). The density is larger in the environment of the nucleus and decreases towards the outer regions of the atom.

Hence, we may consider a crystal lattice as a 3D periodical distribution of electron densities, in the formula

$$F(\vec{h}) = \int_{-\infty}^{\infty} \rho(x,y,z) e^{2\pi i \vec{h}\vec{r}} dV. \tag{3.121}$$

The electron density ρ replaces the former form factor and leads to the above equation which means—considering the e-function as an expression of sine and cosine functions—that a Fourier synthesis of the electron distribution will yield the structure factor for a given **h** (= $h_1h_2h_3$).

Inversely, we may consider the electron density as a Fourier back transformation of structure factors, namely,

$$\rho(\vec{r}) = \int_{-\infty}^{\infty} F(\vec{h}) e^{-2\pi i \vec{h}\vec{r}} dV^*. \tag{3.122}$$

It is evident that we cannot take the Fourier integral, since we only dispose of a non-continuous, limited amount of structure factors. Hence, we have to replace the Fourier integral by a Fourier sum, namely,

$$\rho(x,y,z) = \sum_{h_1}\sum_{h_2}\sum_{h_3} A_{h_1h_2h_3} e^{-2\pi i\left(h_1\frac{x}{a} + h_2\frac{y}{b} + h_3\frac{z}{c}\right)}, \tag{3.123}$$

where $|A_{h1h2h3}|$ = $(1/V_{\text{EZ}}) * |F_{h1h2h3}|$ are the normalized structure factors.

This is the key equation of our method presented here. In order to obtain a good and reliable electron density, it is necessary to dispose of a wide variety of well-measured structure factors together with their phases which stem from a good structure solution (fractional coordinates, space group, lattice constants, a.s.o.) determined beforehand. Otherwise—we will address the problem in detail later in Section 4.3.5—there may arise artefacts: non-real electron densities due to a non-sufficient number of structure factors leading to series-termination errors. For other problems, e.g., the influence of temperature to the measured intensities by the Debye–Waller factor, we refer to the special literature.

From the 1960s, photographical recording methods subsequently were replaced by counter tubes which could be moved to calculated angle positions. It was intended to determine the metrics, space group and atom/ion positions of a given crystal at the same procedure. This was mainly enabled by the great progress in computer technology. Instead of devices with limited possibilities in generating reflections of different layers at the same time, it should be realized to bring each desired plane into reflection position thus covering the maximum of the reflection sphere. It turned out that four angle arcs have been necessary to reach this aim—therefore the related device was denoted as "four circle diffractometer". A schematic view is shown in Fig. 3.41:

Two subsystems may be distinguished at this instrument:

(1) the crystal alignment with ϕ and χ circle
(2) The basis with ω and 2θ circle

The latter ones have a common axis and may be coupled. The "aligner" is, as a whole, attached to the ω-circle and rotates together with it, so that there are two main settings:

- For $\chi = 0°$, the ϕ circle is parallel to ω, 2θ.
- For $\chi = 90°$, the ϕ circle is perpendicular to ω, 2θ.

The counter tube was attached to the 2θ circle in order to put the desired crystal plane into the reflection position and then to scan the reflection in three dimensions to obtain the total intensity resp. count rate by help of the other circles.

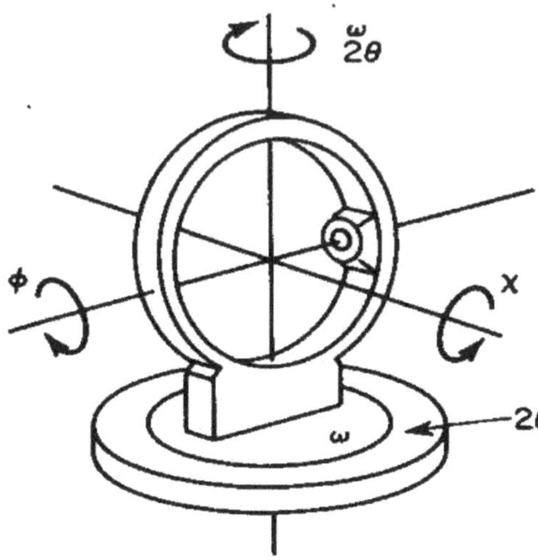

Figure 3.41 Schematic view of a four-circle diffractometer with the "basis" consisting of ω and 2θ circle to turn the crystal into reflection position and the crystal alignment with ϕ and χ circle.

Whereas the ϕ, ω and 2θ arcs were constructed as "full" circles, the χ arc may only be developed partly as a segment ("Eulerian cradle"). A special peculiarity of some producers was the introduction of the κ-circle, where κ was set to a fixed (magic) angle of around 54°, so that a combination of this κ, ϕ and ω replaced the χ circle. This "κ-geometry" or 3-circle instrument has some advantages with respect to the conventional four circle diffractometer (like easier access to the crystal, a higher accuracy and mechanical stability due to the lack of one circle), but also disadvantages (a greater deal of complexity and computer power due to the coupling of different circles to replace the missing one). Later on, the counter tube was replaced by a large CCD chip by which a confined area of reflections could be recorded (3-circle diffractometer)—overlapping parts of the image have to be regarded. A rather recent development on the detection sector are electronic films or image plates, where a comparably large area of reciprocal space can be scanned at the same time. This is in particular interesting for temperature-dependent measurements, where structural phase transitions occur during a narrow time window. These devices commonly need a special layer material where the reflections are recorded in a latent image which is read out by a special LASER scanner. These devices are comparably expensive, but they have the great advantage that structure solutions and other tasks may be performed during a rather short time period.

Lastly, also 2-circle diffractometers should be mentioned here, where the sample has to be pre-oriented with a crystal axis being perpendicular to some circle of the instrument. Such devices mostly were used in cases, where some additional equipment determined the geometry of the diffractometer, i.e. if bath cryostats—these must not be tilted and always have to be kept vertically—belong to the diffractometer. Very often, these instruments are used for neutron scattering.

It is evident that all these types of diffractometers must dispose of utmost mechanical stability and exactness. Turning back to the example of the four-circle diffractometer, the mechanical tolerance, with which the ω (2θ), φ and χ circle turn around a common centre, must be lower than 0.001" (arc seconds!). The focus of the X-ray tube, centre of the crystal aligner and the centre of the detector aperture must lie in a plane exactly perpendicular to the 2θ axis. The aim is that an arbitrary reciprocal lattice point should be brought in contact with the Ewald sphere.

A few words should be dropped on the practice of diffractometry:

(1) The crystal under consideration has to be put on the goniometer head (see above) and the latter one onto the spindle of the diffractometer.

(2) Because of utmost sensitivity the crystal must be placed exactly in the intersection point of the different circles; this can be achieved by help of a combined microscope camera.

(3) This centre of the crystal must not slip out of the crosshair of the camera during the turn of ϕ and χ circle.

    Therefore small crystals (0.2 mm) are best. However, for orientation exposures at large crystals (for purposes of SCMBS), however, the crystals on the relevant camera (very often BPC's) often have to be aimed at the utmost top of the crystal in grazing incidence.

(4) The reflection search is nowadays performed automatically according to a predefined scheme. Having found several

of these reflections, the angle of the crystal and probable lattice parameters have to be evaluated.

These "search reflections" should be strong in intensity, of a high 2θ—value and be distributed among all octants of the reciprocal lattice.

(5) The UB-matrices have to be calculated:

"B" uses quantities to transform *hkl*'s to parameters ($Å^{-1}$) in reciprocal space coordinates.

"U" takes these reciprocal space coordinates and transforms them to internal coordinates of the diffractometer. The control routines calculate the elements of the product matrix UB. Subsequently, cell parameters and their standard deviations are determined.

The result is an orientation matrix in order to find the desired *hkl*'s and to measure their intensity in space.

(6) The measuring program can then be set up and started. For a common structure solution (the main purpose of such diffractometers) the measurement of the different reflections can be rather quick. The determination of the contents of the unit cell (atoms or ions) with fractional coordinates, occupation, overall temperature factors, anisotropic temperature factors can be rather arduous—this depends on the number of formula units within the cell and the size of the unit cell. It may be compared with the solution of a 3D puzzle. However, there are means available (graphical representation of the unit cell, together with effective distances and proposal of atoms/ions) with which the distribution of the particles within the unit cell may be facilitated. For example, very often special ions on certain lattice sites may be ruled out in advance because of their size.

For the purpose of our new method nanoscopy, however, the principles of structure solution are of minor importance. We always have in mind to gather the utmost number of *hkl* structure factors and their phases in order to widely avoid series termination errors. However, also a very well-resolved structure with a minimum *R*-value is mandatory for the evaluation of our electron densities.

3.3.3 Synchrotron Diffraction

Basic concepts for the synchrotron were developed independently from each other by W. I. Weksler in Russia and E. McMillan in Los Alamos, USA. The first electron synchrotron was constructed in 1945 by McMillan, the first proton synchrotron by M. Oliphant in 1952. Later on, after the discovery of the "strong focussing" it was possible to build facilities with energies in the GeV range. A synchrotron is a type of particle accelerator (in the case discussed here: electrons) with the general shape of a hollow ring. A special form of a synchrotron is a storage ring, where the charged particles can be kept after acceleration on a desired energy level. They emit a special sort of X-radiation perpendicular to their direction of motion into the plane of the ring. Strangely enough, in the initial time of the concept and first trial runs, synchrotrons were not meant as being a source of X-radiation; on the contrary, they were planned for nuclear experiments and the occurring radiation was taken as a "nasty waste product", since it led to an energy loss of the particles. However, in the course of time, the very useful properties of their radiation were generally acknowledged, and own electron synchrotrons have been built only for the purpose to use them for diffraction experiments (since the 1970s).

Here a few remarks to the mode of operation of a synchrotron:

It is only at first approximation a ring tube—in detail it consists of deflecting superconductive magnets interrupted by linear acceleration pathways. The acceleration is achieved by high-frequency (HF) fields in empty cavity resonators, the magnets force the electrons on a closed orbit. The tube of the electron "beam" is evacuated (to ultra high vacuum), since the collision of the electrons with some residual air particles would lead to an energy loss. The particles resp. electrons, however, are not accelerated starting with zero velocity: A pre-accelerator or injector forces the particles or electrons (stemming from a glow cathode) to obtain an initial velocity that might be close to c for electrons, and "spits" them into the synchrotron ring. There, the energy and momentum of the electrons will be enhanced, but—according to the requirements of relativistic mechanics—the velocity not. Therefore, in the case of electrons, the synchrotron can be operated in a "quasi-static" mode. Around the ring tube, several devices provide a collimation or amplification of the beam:

First, bending magnets lead to a deflection of the electron "beam" (in reality it is a string of particles) to force it on a circular path whereas the X-radiation is emitted tangentially with a diverging distribution like a fan (Fig. 3.42[9]).

Figure 3.42 General diagram of Synchrotron Soleil. The outer circular ring is the synchrotron, i.e. a particle accelerator that brings electrons (light blue beam) to very high speeds. The electrons are accelerated by electric fields in the straight sections between green squares. The red rectangles are magnets that bend the beam. When the beam is bent the electrons emit synchrotron radiation (shown in yellow), especially X-rays; these are sent into the various beamlines (the straight lines branching out of the synchrotron). Each beamline contains scientific instruments, experiments, etc., and receives an intense beam of radiation.

The height of the fan, however, amounts only to 0.2 mm. Therefore, synchrotron radiation is directed and polarized. This polarization is linear in the plane of the accelerator and elliptical outside. As mentioned above, the "beam" is not continuous, but contains a time structure with pulse periods of 0.1 ms with a pulse distance of 2 ms. However, this is only relevant for very few of the experimental devices, for the other ones the beam might be considered as being continuous. The intensity of this radiation

[9]Copyright © EPSIM 3D/JF Santarelli, Synchrotron Soleil agreed by email to the license below:
The copyright holder of this file allows anyone to use it for any purpose, provided that the copyright holder is properly attributed. Redistribution, derivative work, commercial use, and all other use is permitted.

is expressed as "flux" (with the strange unit (photons per second mrad 0.1% $\Delta E/E$) or brilliance (photons per second mrad^2 0.1% $\Delta E/E$). This brilliance of a beam passing the deflecting magnet, however, is several orders of magnitude larger than that of an X-ray tube (cf. Section 3.3.2) and comprises a factor of 10^4. A further enhancement can be achieved by wigglers (Fig. 3.43) or undulators (Fig. 3.44).

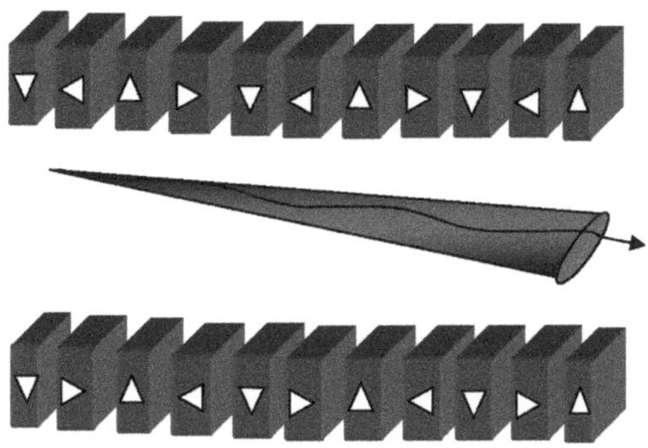

Figure 3.43 Coherent amplification of the synchrotron beam by subsequently arranged poles.

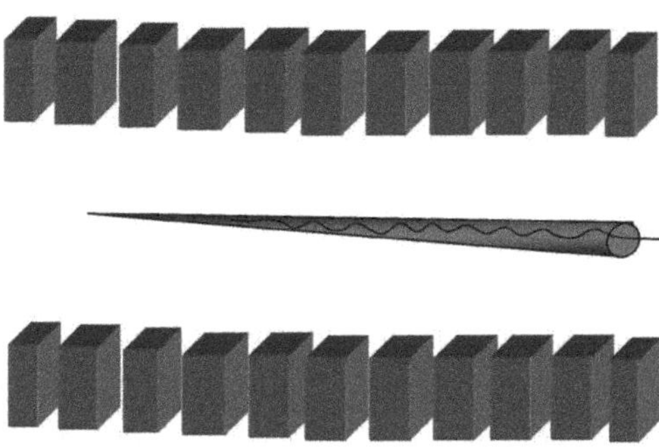

Figure 3.44 Strongly collimated beam with high brilliance by undulators ("wiggler").[10]

The wigglers lead to a long-range oscillation of the beam—the electron packages will be coherently superimposed and thus amplified. The undulators produce short-range beam oscillations which lead to a strong collimation of the beam with very high brilliance. This collimation warrants a minimum divergence and,

[10]Schematic images according to Fuess (2001).

at the same time, a pronounced monochromatization of the synchrotron radiation with a very low wavelength tolerance of $\Delta\lambda/\lambda \approx 5*10^{-4}$. Correspondingly, the resolution of diffraction experiments with this sort of radiation is very high and surpasses the one of X-ray tubes and downstream equipment by far. Generally speaking, the advantages of synchrotron radiation compared to the ones of conventional X-rays are numerous: The sync. radiation is continuous in wavelength and can be "adjusted" to the purposes of the experiment (X-radiation is "characteristic"), the intensity of sync. radiation corresponds to more than 10^{13} photons (X-ray tube: 10^7 photons), the divergence is only 0.02° compared to 15° of an X-ray tube.

For structure solutions a monochromatic, comparably "hard" radiation is chosen, i.e. of an energy range >1 keV = wavelengths shorter than 10 Å. This wavelength can be adapted by the choice of the monochromator or the scattering angle. The scattered radiation is recorded as a function of the scattering angle.

Again, we will leave out a consideration of synchrotron powder patterns—they are characterized by very small line widths and therefore much higher resolution and low overlap than conventional powder diffractograms—and turn towards single crystal diffractometry.

The main advantage here (as compared to conventional X-ray single crystal diffractometry) is the possibility to measure tiny crystals of a few μm^3. Sometimes only these small specimen are available—in particular the synthesis of several compounds does not allow larger crystals. However, this tiny crystal must be of very good quality—it is a widespread error to think that the outstanding properties of synchrotron radiation might "cure" everything—a bad (i.e. twinned or highly distorted) crystal remains a bad crystal. In some cases, a tiny, well-grown single crystal as part of a large bad crystal may be etched and, contrary to conventional X-ray diffractometry, used in a synchrotron experiment, but this is another case as the one mentioned above.

However, a real advantage due to the high brilliance and allowance for tiny crystals, is the minimizing of effects like absorption and extinction—sometimes a severe obstacle to a determination of accurate intensities at conventional X-ray diffractometers. The same is achieved by the application of very short wavelengths (0.5 Å), so that the accuracy of electron densities could be enhanced considerably. An example is shown in Fig. 3.45.

Figure 3.45 Static electron density of kalifeldspar sanidine within two planes from synchrotron diffraction measurements.[11]

This image is similar to the numerous two-dimensional drawings of electron densities ("contour maps") from late 1970s/early 1980s being nowadays still widespread in the crystallographic community. The resolution of these maps could be enhanced considerably by use of synchrotron radiation—up to the point of multipole refinement: Sometimes the resolution of these maps is so excellent, that not only the normal form factor resp. spherical symmetrically based refinement can be performed but also an overlay with, e.g., d-orbitals. An example is shown in Fig. 3.46.

It was applied by the co-author of several of our recent publications, Armin Kirfel and T. Lippmann (2002), for the case of synthetic fayalite. The multipole refinement (recognizable at the high-density orbital contours around the site of the nuclei) of their synchrotron data of one of our crystals led to an R-value of only 0.6%(!). The map may be considered as the ancestor of our three-dimensional difference electron density plots (cf. Chapter 5) by which the most important quantity mentioned in the present book, the electric field gradient, was calculated. For these plots a multitude of several thousand structure factors had to be evaluated by Fourier back-transformation. It will be described below in detail how this was achieved.

[11]Images according to Fuess (2001).

Figure 3.46 Static deformation density (difference electron density) of synthetic fayalite in the planes through O(2)-M2-O(3), contours at 0.05 e/Å^3.

3.3.4 Neutron Diffraction (Nuclear and Magnetic)

Neutrons are electrically neutral particles (therefore the notation) in the atomic nucleus and were first detected by Chadwick (Nobel Prize 1935, cf. Section 3.3.2.5). According to Wolfgang Pauli, the neutron decays with a half lifetime of 10.6 min. into a proton, an electron and an antineutrino according to the following equation:

$$n^0 \rightarrow p^+ + e^- + \nu^- \quad (3.124)$$

Since neutrons n^0 do not bear a charge, the detection and counting of scattered neutrons (in order to measure an intensity) have to be performed by a secondary reaction, namely,

$$^3He + n^0 \rightarrow {}^3H + p^+. \quad (3.125)$$

The charged proton will cause a count pulse in a proportional counter and therefore is equivalent to an incoming neutron. The multitude of these neutrons are then a measure for the recorded intensity of, e.g., a certain reflection.

How are these neutrons produced? There are reactions in the laboratory scale, but these are by far too weak to serve as a source for neutron diffraction purposes. Suited for the latter are neutrons which are produced in a nuclear reactor or spallation source. The latter possibility was evaluated rather late: Accelerated heavy particles like protons hit a target of special nuclei and transform the latter in an excited state. By "evaporation" of particles (in particular neutrons) the excited nuclei will drop back into the ground state—the neutrons are "spat" out ("spallated"). Such sources produce a high flux of neutrons, but they have a time structure—they are rather rare and can be found at Los Alamos, Argonne (both USA) or at Rutherford Appleton Lab at Gt. Britain.

The former possibility is the commonly applied method: The neutrons are produced by fission of the uranium isotope ^{235}U according to the following equation:

$$^{235}U + n^0_{\text{therm}} \rightarrow {}^{236}U \rightarrow X + Y + \nu \cdot n^0_{\text{fast}} \tag{3.126}$$

X, *Y* shall denote any fission products. The primarily generated fast neutrons have a special dependence on energy, which resembles a Maxwell distribution with a maximum at approximately 1 MeV. In order to operate a continuously working reactor two conditions have to be met:

(1) The reproduction rate of neutrons ν must be larger than 1 (normally $\nu = 2.5$).
(2) The produced fast neutrons with high energy do not lead to further fission reactions—therefore they have to be braked to lower velocities in the thermal energy range by a suitable moderator. Commonly, this is D_2O or graphite—the uranium containing rods point into a "bathtube" with D_2O—by this measure the maximum of the Maxwell distribution is shifted to lower ("thermal") energies. The temperature of the "bath" is led off by a secondary cooling circuit, the usable neutrons by special beam tubes. By certain containers in the reactor core with liquid hydrogen (T = 22 K, "cold source") or heated graphite (T = 2200 K, "hot source"), the energy spectrum of the neutrons can be enhanced considerably—the range of available wavelengths amounts to $0.5 \text{ Å} < \lambda < 20 \text{ Å}$.

The research reactors are optimized for the production of neutrons, therefore they are not characterized by the quantity "power" but by the notation "flux" (neutrons/cm^{2}s in the core). The best reactors of this kind attain values of 2×10^{15} n^0/cm^{2}s (e.g., Brookhaven, USA or Grenoble, France). However, only part of this flux (namely 10^8 n^0/cm^2s) really hits the sample after monochromatization. The latter is achieved mostly by crystals in a certain angle position (e.g., Cu(200), Ge(111), Si(111)) or by "choppers" (rotating slit wheels).

The beam tubes point into the core (perhaps towards a "hot" or "cold" source) and lead part of the neutron flux to the different instruments. These are arranged in such a manner that the space around the well-screened core, the reactor hall, is optimally utilized under the conditions of possible flux, wavelength, energy a.s.o.—therefore the beam tubes are by no means arranged like the spikes of a hedgehog, but more complicated.

However, now a few words to the neutrons themselves with their most valuable properties:

(1) Since neutrons bear no charge, there is no Coulomb interaction with charged centres of scattering—therefore the depth of impression into matter is huge.
(2) The neutron, since being a nuclear particle, is mainly scattered at nuclei. Contrary to X-rays with their form factor dependence of scattering power, the nuclear scattering of neutrons is dependent on the connecting conditions between the nuclear wavefunction and the wavefunction of the neutron—hence, it may be expressed only by a parameter with the dimension "length", the so-called scattering length.
(3) In spite of charge neutrality, the neutron has an inner structure and a magnetic moment—therefore it can be scattered at unpaired magnetic electrons.
(4) The mass of the neutron (1.0087 units) is almost the same as the proton mass (1.0073 u). 1 u = 1.66×10^{-27} kg.
(5) The wavelength of thermal neutrons is comparable with the distances of atoms/ions in a crystal lattice—therefore it is possible to perform structure investigations yielding the site of the nuclei.
(6) The energy transfer of thermal neutrons is of the same order as their initial energy—therefore scattering at phonons (dynamical processes) is measurable.
(7) The decay of neutrons already has been mentioned above.

Let us consider points 2 and 3 in detail.

Since the nucleus with its diameter in the order of 10^{-15} m = 1 fm is much smaller than the wavelength of neutrons, no change in the scattering power as a function of the scattering angle can be observed. The scattering power of the nucleus is a constant, denoted by *b* with the unit fm. There are exceptions if the wavelength of the neutrons is situated in the neighbourhood of a nuclear resonance—this is called anormal dispersion.

Since the configuration of the atomic nuclei is strongly different for even neighbouring nuclei and isotopes, the change depending on the atomic mass is discontinuous. Even isotopes of the same element may dispose of different scattering lengths. At present, the theoretical models for this discontinuous behaviour of the scattering lengths, however, are not sufficiently developed and therefore theoretical predictions of *b* not very exact. The values of the tables have been determined experimentally. A schematic view of the properties is given in Fig. 3.47:

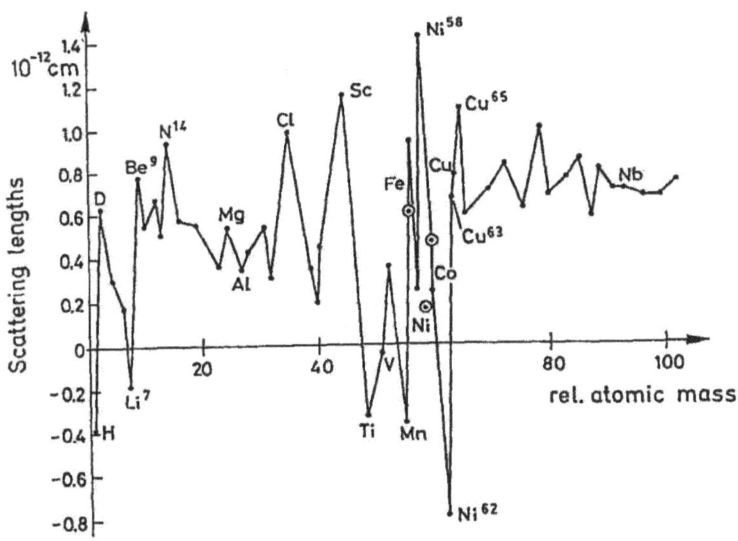

Figure 3.47 Scattering lengths for neutrons of elements and isotopes.[12]

The extremely discontinuous change of the scattering lengths renders it possible to determine the distribution of neighbouring cations in a crystal lattice. For example, the difference between *b*(Fe) = +9.54 fm and *b*(Mn) = −3.73 fm is so severe that, e.g., olivines with the same structure, but different cation distributions can be easily distinguished in neutron powder diffractograms. Figures 3.48–3.50 show the latter for the three olivines Fe_2SiO_4, $FeMnSiO_4$ and Mn_2SiO_4, respectively:

[12]Diagram according to Fuess (2001).

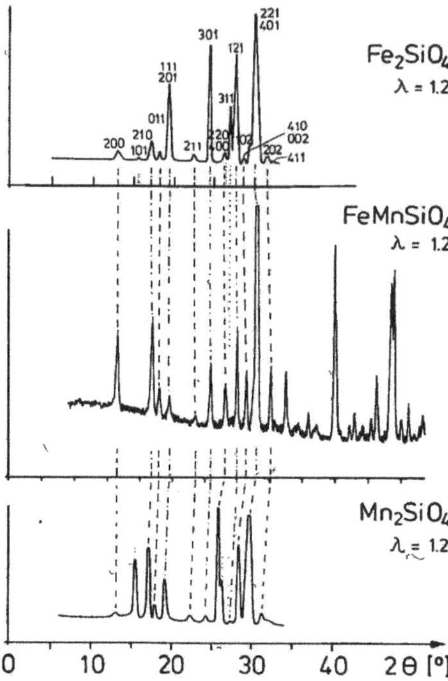

Figures 3.48–3.50 Neutron powder diffractograms of the olivines Fe_2SiO_4 (fayalite), $FeMnSiO_4$ (fayalite-tephroite) and Mn_2SiO_4 (tephroite).[13]

The peaks with the same indexing are connected by dashed lines. Hence, it may easily been noticed that the different cation distributions on the two possible crystallographic sites M1 and M2 lead via the different scattering lengths to patterns which can well be distinguished from each other.

Corresponding X-ray patterns due to the almost equal form factors of Fe and Mn would never allow to distinguish between the two cations.

Another nice difference in scattering lengths is valid for hydrogen $b(\mathrm{H}) = -3.74$ fm and deuterium $b(\mathrm{D}) = 6.67$ fm. By that, it is possible to even find water molecules in a structure. "Deuteration", the exchange between hydrogen and deuterium, allows to distinguish between differently bound water molecules. This is in particular interesting for zeolites, where water molecules are often attached to certain ions in the zeolite cages—the absorbing and desorbing of water molecules (or other ones) is

[13] Images according to Fuess (2001).

eminently important for practical applications of the relevant zeolites concerning molecular sieves. Also, the location of hydrogen bridge bonding can be performed by neutron diffraction measurements.

A very nice and quite popular application are the so-called self-cooling beer containers. These devices contain a double wall in which a special zeolite under vacuum and a separate water reservoir are located. Just by opening of a valve, the water will be adiabatically distributed in the giant inner surface of the zeolite (due to the cages within the structure) and the energy for it taken from the reservoir: The water may even freeze during this reaction.

Hence, the container and the beer will cool down without using an external device (like a refrigerator). After the enjoying of the cool beer, the container will be given back to the brewery, the container will be refilled and the double wall again be evacuated—so the process can start a further time.

A more scientific result was the determination of the cation position in the above-mentioned olivines: The occupation of the M1 position by Fe^{2+} cations is preferred, whereas the M2 position is dominated by Mn^{2+} ions.

Another problem of great interest in mineralogy is the distribution of silicon and aluminium in the basic units of silicate minerals, the tetrahedra. By the different scattering lengths of Si and Al, a distinguishing of these two cations (in particular in feldspars) is possible.

However, let us now consider the nuclear scattering in detail:

A parallel stream of neutrons of density j and wave vector $\mathbf{k}$ might hit a thin target with the area A with $a*A$ $(a < 1)$ scattering centres with single area σ. Under the angle θ with respect to the direction of incidence, a detector will record (count) the neutrons with wave vector $\mathbf{k}'$ being scattered into the space angle element dΩ. A very common measure to describe nuclear scattering processes is the differential cross section. It is defined as follows:

$$\left(\frac{d\sigma}{d\Omega}\right)_\theta = \frac{\text{number of neutrons/time scattered into } d\Omega}{\text{stream density of the incident neutrons}} \tag{3.127}$$

Coherent scattering is only found for constructive interference of the scattered waves, expressed by the exponential function in the following equation:

$$\left(\frac{d\sigma}{d\Omega}\right)_{\text{coh}} = \left|\bar{b}\right|^2 \left|\sum_j e^{i\vec{q}\vec{r}_j}\right|^2 \tag{3.128}$$

Therefore the mean scattering length b and a term describing the site of the nuclei within the crystal is decisive for the differential σ. Deviations from this mean scattering potential will lead to incoherent scattering amounts which will not further be regarded.

We will now consider a crystal with atoms on the sites r_j with a spatially periodic distribution of unit cells. Above, we have defined the construction of a unit cell by the basic lattice vectors **a**, **b**, **c** and the volume $V_0 = \mathbf{a}(\mathbf{b} \times \mathbf{c})$. Then we get a lattice vector n with the components:

$$\mathbf{n} = n_1\mathbf{a} + n_2\mathbf{b} + n_3\mathbf{c} \tag{3.129}$$

If the unit cell contains only one crystallographic site, we will get a Bravais lattice, in the general case of several atoms or sorts of atoms per unit cell, we have to construct the cell from several Bravais lattices penetrating each other (cf. Lottermoser, 2015). The atomic resp. nuclear positions will then be denoted by

$$\mathbf{r}_d = x_d\mathbf{a} + y_d\mathbf{b} + z_d\mathbf{c}, \tag{3.130}$$

with $0 \le x, y, z \le 1$ and d denoting the d-th sort of atoms.

As mentioned above, the reciprocal lattice is the key property of scattering experiments with the reciprocal lattice vector

$$\mathbf{g} = h\mathbf{a}^* + k\mathbf{b}^* + l\mathbf{c}^* \tag{3.131}$$

with, e.g., $a^* = 2\pi\frac{\vec{b}\times\vec{c}}{\vec{a}(\vec{b}\times\vec{c})}$; the other components being cyclic permutations of the three lattice vectors **c** and **a** and **a** and **b**.

Our equation from above, taken with the scattering vector **q** and the lattice vector **n** will give

$$\left(\frac{d\sigma}{d\Omega}\right)_{\text{coh}} = \left|\bar{b}\right|^2 \left|\sum_n e^{i\vec{q}\vec{n}}\right|^2 \quad \text{(for a Bravais lattice)} \tag{3.132}$$

If **q** is equal to 0 or a reciprocal lattice vector, the right hand side of the above equation will become huge, because the terms under the sum will add up. If **q** is not equal to a reciprocal lattice vector, the terms are not coherent and will not obey Bragg's law. We may express these properties by

$$\left|\sum_n e^{i\vec{q}\vec{n}}\right|^2 = \sum_g \delta(\vec{q}-\vec{g})N\frac{(2\pi)^3}{V_0} \tag{3.133}$$

The last term was obtained by integration over dq in the unit cell. Hence, we get for unit cells with several sorts of atoms:

$$\left(d\sigma/d\Omega\right)_{\text{coh}} = N\frac{(2\pi)^3}{V_0}\sum_g \delta(\vec{q}-\vec{g})\left|F_N(\vec{g})\right|^2 \tag{3.134}$$

The δ-function expresses the Bragg condition. The nuclear structure factor is

$$F_N(\vec{g}) = \sum_{\vec{r}_d} e^{i\vec{g}\vec{r}_d}\overline{b}_{\vec{r}_d} \tag{3.135}$$

These properties can be found in the famous books of Bacon (1975) and Marshall-Lovesey (1971).

Summarizing, we may express a nuclear structure factor by the scattering length of a given nucleus and by its position within the unit cell multiplied by a reciprocal lattice vector, so that we obtain

$$F_N(\vec{g}) = \sum_d^s \overline{b}_d\, e^{2\pi i(x_d h + y_d k + z_d l)} \tag{3.136}$$

The inner products **aa***, **bb***, **cc*** will give unity, the other terms zero, so that the above expression remains.

As already mentioned under point 3 of our introducing remarks (Section 3.3.4), neutrons can be scattered by magnetic moments of electrons belonging to partly occupied outer shells (3d, 4d, 4f, 5f). The magnetic scattering is superimposed by nuclear scattering. In paramagnetic substances (non-spontaneously ordered magnetic moments) only the background will be enhanced. In the case of collective magnetic phenomena (ferro-, ferri-, antiferromagnetism), magnetic scattering leads to Bragg reflections. These cause in the case of antiferromagnetic ordering superstructure reflections, whereas in case of ferromagnetic ordering the periodicity of nuclear and magnetic cell is identical—the magnetic scattering adds up to the nuclear ones at the same angles. Whereas the scattering length is independent on these scattering angles, the magnetic contribution is characterized by a form factor corresponding to the one of X-ray diffraction.

There are many publications concerning magnetic structures and magnetic phase transitions, e.g., for natural and synthetic garnets (Hock, 1990; Hock et al., 1991), olivines (Fuess et al., 1988) and other. Many of the studies have been performed by Rietveld analysis of powder samples. A nice example is again shown for the case of our above-mentioned three olivines fayalite Fe_2SiO_4, fayalite-tephroite $FeMnSiO_4$ and tephroite Mn_2SiO_4 (Figs. 3.51–3.53).

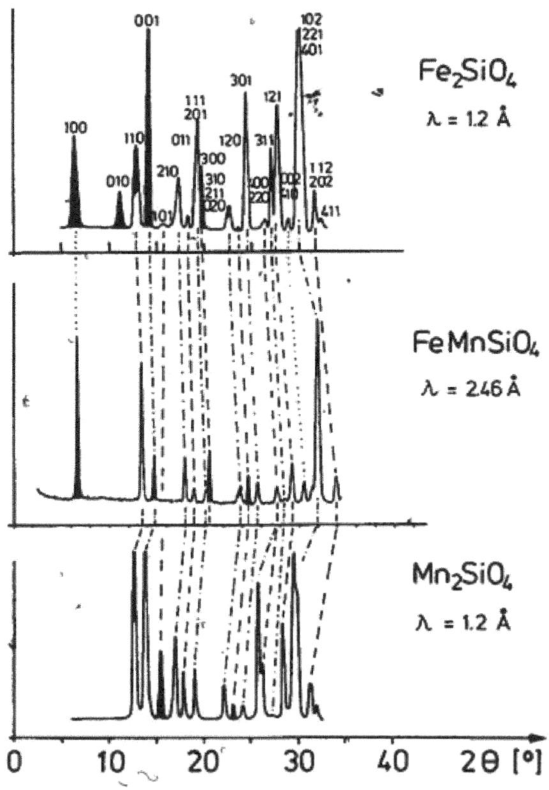

Figures 3.51–3.53 Powder diffractograms of the three olivines fayalite Fe_2SiO_4, fayalite-tephroite $FeMnSiO_4$ and tephroite Mn_2SiO_4 in the antiferromagnetic state.[14]

Again, the dashed lines show the analogous peaks of the neutron powder patterns; this time the black coloured ones denote the

[14]Images according to Fuess (2001).

magnetic contributions—the differences in the magnetic structures of Fe_2SiO_4 and Mn_2SiO_4 arise due to different antiferromagnetic alignments (canting).

Again, let us consider the magnetic scattering of neutrons at unpaired magnetic moments a bit closer:
We also start from an expression for the differential σ:

$$\left(\frac{d\sigma}{d\Omega}\right)_{\text{magn}} = \left(\frac{\gamma e^2}{m_e c^2}\right)^2 \left|\int d\vec{r}\, e^{i\vec{q}\vec{r}} (\vec{q}\,' \times \rho_S(\vec{r}) \times \vec{q}\,')\right|^2 \qquad (3.137)$$

This equation is a bit similar to the analogous terms for the nuclear differential σ, besides the double cross product in brackets. ρ_S denotes the fact that the wavefunctions of the unpaired electrons are not localized at the sites of the nuclei, but we really have to consider spin densities $\rho_S(r)$ instead of localized spins s (orbital terms are neglected, as mostly the case for 3d magnetism). This spin density should be periodic, therefore we may replace this term by

$$\int_{\text{unit cell}} d\vec{r}\, e^{i\vec{q}\vec{r}}\, \rho_S(\vec{r}) = \frac{(2\pi)^3}{V_0} \sum_{\vec{g}} \delta(\vec{q} - \vec{g}) \vec{F}(\vec{g}) \qquad (3.138)$$

where **F(g)** is the vector magnetic structure factor and can be re-transformed into a spin density by Fourier inversion (or equivalent: Fourier back-transformation):

$$\rho_S(\vec{r}) = \frac{1}{V_0} \sum_{\vec{g}} e^{-i\vec{g}\vec{r}}\, \vec{F}(\vec{g}) \qquad (3.139)$$

This equation is the key of our later 3D reconstructions of magnetic electrons in the unit cell of synthetic fayalite. Inserting the last but one formula into our main expression for the differential σ, we get

$$\left(\frac{d\sigma}{d\Omega}\right)_{\text{magn}} = \left(\frac{\gamma e^2}{m_e c^2}\right)^2 \frac{N(2\pi)^3}{V_0} \sum_{\vec{g}} \delta(\vec{q} - \vec{g}) \left|(\vec{q}\,' \times \rho_S(\vec{r}) \times \vec{q}\,')\right|^2 \qquad (3.140)$$

The δ-function again denotes the Bragg condition, g the gyromagnetic ratio, e the electron charge and $m_e c^2$ Einstein's formula for the energy of an electron. The double cross product will be explained later.

In the special case for a simple antiferromagnet with identical sublattices (single-domain case) we obtain

$$\left(d\sigma/d\Omega\right)_{\text{magn}} = \left(\gamma e^2/m_e c^2\right)^2 \frac{N(2\pi)^3}{V_0} \sum_{\vec{g}} \delta(\vec{q} - \vec{q}) \left|\vec{F}_M(\vec{g})\right|^2 (1 - (\vec{g}'\vec{e})^2) \quad (3.141)$$

where $\mathbf{F}_M$ is the magnetic structure factor with

$$\vec{F}_M(\vec{g}) = \sum_d \frac{1}{2} g_d < \hat{S}_d > f_d(\vec{g})(\text{sign})\, e^{i\vec{g}\vec{r}_d} \quad (3.142)$$

Here, the general case of a unit cell with several atomic sorts have been considered—this is denoted by the dependence on d. f_d is the atomic form factor being 1 for $\mathbf{q} = 0$ and decreasing according to a Bessel function dependent on $\mathbf{q}$. "sign" is the sign of the spin direction of the relevant sublattice. Analogously to the differential σ without magnetic contribution, we will get in this elastic-coherent case only contributions for $\mathbf{g} = \mathbf{q} = \mathbf{k} - \mathbf{k}'$ (δ-function!) under obedience of the Bragg law.

The magnetic structure factor $\mathbf{F}_M$ contains like the nuclear one a phase contribution $\exp(i\mathbf{g}\mathbf{r}_d)$, which regards the distribution of atoms within the unit cell. In contrast with F_N we have a dependence on the expectation value of the spin operator <S> which is proportional to the magnetic moment. Again in contrast with the nuclear case we obtain a modulation of F_M via the atomic form factor f_d. The latter depends through $\mathbf{q}$ or $\mathbf{g}$, respectively, on sineθ/λ, according to Bragg's law. This form factor describes the distribution of the magnetic electrons (in our case: 3d electrons) around the nucleus analogously to the form factor of X-ray diffraction.

The slope of the latter, however, with sineθ/λ will be less pronounced since d-electrons are more distant from the nucleus than the total electron distribution in the X-ray case. Provided a non-complete quenching of the orbital contribution (the common case for 3d electrons, one mostly observes pure spin magnetism!); this effect can be regarded by another Bessel function in the form factor decrease. Similar to the dependence of the form factor on sineθ/λ covalency effects are acting which will be considered later in Chapter 5. As we have seen before, the nuclear structure factor does not contain a dependence on a form factor, but solely on a scattering length.

The last term for the differential σ takes the observation into account that the neutrons do not "see" the normal magnetic structure factor $\mathbf{F}_\mathrm{M}$ itself, but only its projection $F_\mathrm{M}^{\perp}$ onto the scattering plane—this is denoted by the double cross product mentioned above.

As a correction factor for the "real" magnetic structure factor we introduce the coefficient q''^2, since we already have used scattering vector $\mathbf{q}$ and normalized scattering vector $\mathbf{q}'$. Hence, we get

$$\left|\vec{F}_M^{\perp}\right|^2 = q''^2 \left|\vec{F}_M\right|^2 = \left(1-(\vec{q}\,'\vec{e})^2\right)\left|\vec{F}_M\right|^2 \xrightarrow{\text{Bragg}} \left(1-(\vec{q}\,'\vec{e})^2\right)\left|\vec{F}_M\right|^2 \quad (3.143)$$

where $\mathbf{q}'$ is a unit vector in the direction of the scattering vector and defined as $\mathbf{q}/|\mathbf{q}|$, $\mathbf{e}$ is a unit vector denoting the spin direction with

$$<\hat{\vec{S}}> = <\hat{S}>\vec{e} \quad (3.144)$$

Correctly in the general case (not in the special case of a single-domain linear antiferromagnet as mentioned above) the vector property of the magnetic structure factor $\mathbf{F}_\mathrm{M}$ has to be taken into account. In our case of synthetic fayalite at low temperatures (spin canting on the M1 site), however, the components of the magnetic moment with respect to the crystallographic axes μ_a, μ_b, μ_c have to be regarded by using a column vector.

In principle we now dispose of the tools to understand the procedure of evaluation of a magnetic structure by neutron scattering.

3.3.4.1 The diffracted intensities

In the beginning of the sub-chapter it was mentioned that the scattered intensity be proportional to the differential σ and hence, to the square of the structure factor. How this intensity is determined through the nuclear and magnetic structure factor, however, depends on the fact, whether polarized or unpolarized neutrons have been used. As being shown already in 1939 by

Halpern and Johnson (1939), nuclear and magnetic structure factor are superimposed in case of complete polarization of the incident beam according to the following equation:

$$I_{\text{pol}} \propto \left(F_N \pm \vec{F}_M^{\perp} \right)^2 = F_N^{\,2} \pm 2F_N \vec{F}_M^{\perp} + \vec{F}_M^{\perp 2} \tag{3.145}$$

In the case of completely unpolarized (or "integrated") neutrons, however, the interference term in the above equation vanishes and we obtain

$$I_{\text{int}} \propto F_N^{\,2} + \vec{F}_M^{\perp 2} = I_N + I_M^{\perp} \tag{3.146}$$

In this case the single structure factors do not superimpose like in the polarized case, but the single intensities I_N and $I_M^{\perp}$ do so. By this, the magnetic contribution can be easily separated provided the nuclear structure has been determined before.

The special advantages of polarized neutrons, however, can be verified by closer inspection of the observable of spin-polarized neutrons, the so-called flipping ratio R. This is defined as the ratio of the intensities with negative and positive spin direction with respect to the quantization axis:

$$R = \frac{I^-}{I^+} \propto \frac{(F_N - \vec{F}_M^{\perp})^2}{(F_N + \vec{F}_M^{\perp})^2} = \frac{(1-\gamma)^2}{(1+\gamma)^2} \quad \text{with}\ \gamma = \frac{\vec{F}_M^{\perp}}{F_N} \tag{3.147}$$

For this consideration, let us assume that the magnetic structure factor amounts to 1% of a normalized nuclear structure factor. From the above equation we may conclude that the contribution of $\vec{F}_M^{\perp}$ to the measured flipping ratio amounts to ≈4%. Concerning the integrated (unpolarized) neutrons, however, the contribution of the magnetic part to the total intensity amounts to only 0.1‰. Hence, the method of using spin-polarized neutrons is an excellent tool to determine very low magnetic structure factors which may be detected at high values of sine θ/λ (at the "outskirts" of the form factor slope). Generally speaking, it is even possible to derive experimental form factor curves of the ion involved. Moreover, it is possible (via the equation mentioned above)—by Fourier inversion of the exactly determined magnetic structure factors—to derive a spin density map of the unit cell of the crystal under consideration. The former is just the local distribution of magnetic moments in the direct space. In this case, however, a very large number of flipping ratios has to be measured.

However, there is one general issue: Via the coefficient γ, the ratio of magnetic and nuclear structure factor, the accuracy of $F_M^\perp$ is strongly dependent on the accuracy of F_N. This means that the nuclear (i.e. the crystallographic) structure should have been determined with an excellent accuracy (low value of the goodness-of-fit-parameter) before starting a polarized neutron experiment. In particular for low values of F_N, the connected value of $F_M^\perp$ will become extremely unsecure.

A further disadvantage is the fact that the crystal acts analogously to the analyzer in the case of optical polarization measurements: The incident neutrons will "see" only the component of $F_M^\perp$ parallel to their polarization vector; hence, the polarized neutron experiments are commonly performed by aligning the "easy" axis (direction of spontaneous magnetization) parallel to the spin direction of the incident neutrons—only in case of collinear ferro- and antiferromagnetic substances, best results are then achieved.

Lastly, only with centrosymmetric crystals the determination of the magnetic structure factors can be solved at limited expense, with non-centrosymmetric structures one will get difficulties during the evaluation of phases.

The mentioned disadvantages, however, are surpassed by the benefits by far. A further advantage concerns the necessary corrections of the measured quantities.

3.3.4.2 Corrections of observables

The nuclear scattering factor and the components of the magnetic structure factor are—in principle—numbers, whereas the measured reflections show profiles which may be approximated by a Gauss distribution. In order to establish a connection between the (ideal) structure factors and the observed intensity, we have to apply a multitude of sample- and device- specific corrections and data reductions.

The latter are mostly performed by special program systems just after the measurement. For example, at Grenoble the reflection profiles (which have accurately been centred before) are followed in a step-scan technique with discretization of the peak in dependence on the scattering angle: The measured count rate at each step is referred to the count rate of the incident neutrons in order to get a relative measure for the intensity. Afterwards a correction for background is applied and the integrated intensity reduced concerning Lorentz contribution—this takes into account that net planes stay in reflection position for different times; this leads, due to different counting times, to an artificial hierarchy of reflection intensities which has to be corrected.

3.3.4.3 Sample specific corrections (non-standard reductions)

(a) Temperature factor

Here, the above-mentioned Debye–Waller factor has to be considered. Instead of being fixed at special sites in the crystal lattice, the atoms perform thermal oscillations which lead to an additional phonon scattering of the incident neutrons and thus reduce the elastic coherent differential σ. The necessary correction will be applied as a multiplicative factor directly to the nuclear structure factor:

$$F_N(\vec{q}) = \sum_d \bar{b}_d \, e^{i\vec{q}\vec{r}_d} e^{-W_d(\vec{q})} \quad \text{... with } e^{-W_d} = \langle e^{-\vec{q}\hat{\vec{u}}} \rangle \tag{3.148}$$

The operator $\hat{\vec{u}}$ describes the dilatation of the considered atom or ion. In harmonic approximation we get

$$W(\vec{q}) = \tfrac{1}{2} \langle (\vec{q}\hat{\vec{u}}) \rangle \tag{3.149}$$

The Debye–Waller factor is, via Bragg's law, dependent on $(\text{sine}\,\theta/\lambda)^2$; the dilatation is regarded as a fit quantity. However, in the general case, the temperature oscillations are anisotropic, since the close environment and hence, the couplings of the next neighbours of the considered atom/ion, will be different for different directions in space. The temperature factor is then a tensor and is regarded by a refined table of u-components.

(b) Scale factor

With this parameter, an overall proportional factor between observed and calculated intensities is introduced; it contains the constant peak-background-ratio dependent on different neutron sources—therefore, not the absolute, but the relative intensities are decisive for the structure solution.

This is different in the case of polarized neutrons: The experimental magnetic structure factor is obtained in absolute values in units μ_B. The calculated magnetic structure factor is given in the same units, a scale factor is obsolete.

(c) Absorption correction

By this term it is taken into account, that not only a scattering differential cross section may be effective but also an absorption differential one. In general, the following equation is valid:

$$I = I_0 \, e^{-\mu x}, \tag{3.150}$$

where μ is the linear absorption coefficient and x the transmitted sample thickness.

For a given reflection the absorption is dependent on the path length that the neutron beam has to absolve within the crystal; thus it also depends on the Bragg angle. For this, the dimensions of the sample have to be known, since with the usually applied sample plates with their small edge parallel to the incident beam, the absorption will be low in these directions, but huge in directions perpendicular to the latter. To account for this effect, programs have been developed in order to calculate the effective path lengths for each reflection *hkl* and thus to correct the observed intensity.

(d) Extinction correction

This phenomenon is not so well established as the preceding ones, but there are basic approaches for the quantitative solution of the problem. In general, it is distinguished between primary and secondary extinction.

Concerning the primary extinction one assumes an ideal single crystal which is transmitted by a coherent wave field. By this event, interference effects between incident and scattered waves might occur according to the "dynamic model"; both are thus coupled to each other by amplitude interaction. The bigger the size of the crystal—a necessary property at neutron scattering—the more the interference effects will be enhanced.

The secondary extinction is described according to the "kinematic" theory of diffraction: The crystal is taken as a multitude of microscopic, slightly tilted mosaic blocks. Within the latter the primary extinction can be neglected; in this case the scattered intensities are superimposed (intensity coupling). This model has been evaluated by Zachariasen (1967) and Becker and Coppens (1974) quantitatively.

Actually, both cases cannot strictly be separated—it is possible, however, to assume a crystal with mosaic blocks, whose primary extinction cannot be neglected. For a coefficient Y which connects the calculated intensity with the experimental one, we obtain,

$$I_{\text{exp}} = I_{\text{cal}}\, Y, \tag{3.151}$$

with $Y = 1/\sqrt{1 + 2x + A(\theta)x^2/(1 + B(\theta)x)}$, where A and B are functions of the scattering angle.

The parameter x contains different quantities; the mean path length within the crystal, intensity, wavelength, volume of the unit cell, diffraction angle θ and a further parameter which is—depending on the distribution function of the mosaic blocks (according to Gauss or Lorentz)—differently determined by a quantity r (domain radius) and g (mosaic block distribution). In some refinement programs these two quantities are the fitting parameters while the other ones are calculated from experimental and sample-specific properties. For the experiments with spin-polarized neutrons the extinction differently acts on the "spin-up" intensities I^+ and the "spin-down" intensities I^-, respectively. Whereas the above "geometrical corrections" cancel each other on average, this is not valid in the case of extinction and one has to introduce two extinction corrections Y^+ and Y^- which are calculated according to the formalism of Becker and Coppens (1974).

Strangely enough, the above effect occurs mainly within "perfect" single crystals. Whereas the latter, from reasons of a good structure refinement, should normally be preferred, they often display huge extinction effects due to rather undisturbed net planes and absence of mosaic blocks. This may even lead to a complete failure of structure refinement. In this situation an old trick among neutron diffraction users consists of "tempering" such a crystal by putting it into liquid nitrogen. This may be troublesome (the crystal might crack) and the author rejects all claims arising from this advice, but in some cases the mosaicity of the crystal could be enhanced and the extinction effects diminished.

(e) Magnetic domains

This effect only concerns measurements with spin-polarized neutrons. In some cases the occurrence of antiphase domains in a collinear antiferromagnet with a distribution of 50:50 for the one or other sort leads to a complete disappearance of the magnetic scattering contribution (the flipping ratio approaching unity). Using a strong magnetic field, however, will shift the equally distributed domains towards the preference of one sort, e.g. 70:30. In our measurements concerning tephroite Mn_2SiO_4 we managed to remove antiphase domains with a maximum field of 4.6 T, whereas in the case of fayalite such trials were in vain.

(f) Device specific corrections

These are due to non-ideal behaviour of the applied diffractometer and its influence to the observable. One such type is the so-called $\lambda/2$-contamination. For the monochromatization of the primary neutron beam from the reactor core, special crystals are used, which are brought into reflection position for a special *hkl* triple. By inversion of the Bragg equation, a certain wavelength will be segregated. This new beam will then be taken as the new primary beam for the experiments.

Under this angle position of the monochromator crystal, however, other net planes parallel to the first one are in Bragg position (with whole multiples of *hkl*)—this leads to a contamination of the primary neutron intensity with higher harmonics of the basic wavelength—mainly $\lambda/2$. Normally, this effect can be neglected. However, in the case of polarized neutrons, this phenomenon again affects I^+ and I^- differently and should be regarded.

Another issue here is the grade of polarization: This may be different from unity (perfect polarization: 100%); the effect of the so-called cryoflipper can be different from unity as well—this leads to a non-complete spin flip.

However, these are problems which are monitored carefully and taken into account.

3.3.4.4 The neutron diffractometers

Our experiments on synthetic fayalite (and other compounds) were performed at the instrument D15 at the Institute Laue-Langevin at Grenoble. The device was constructed similarly to the schematic image of Fig. 3.41, but the equipment (a ^{4}He bath cryostat) required a modification of the circles involved, because a tilt of the cryostat was not allowed. The detector (a ^{3}He counter) moves on the χ-circle, which is constructed only as a short arc, and on the 2θ-circle. The cryostat turns together with the sample on the ω-circle—the ϕ-circle was left out. Therefore the observable directions in space are limited—the crystal has to be mounted with a pre-orientation on the sample holder. In our case, a special crystal axis had to be found by previous X-ray diffraction and the crystal glued with this axis being vertical on the sample holder, so that the two other crystallographic axes could be measured perpendicularly to it (2-circle geometry). As a monochromator, a Cu-crystal in 331-position has been used, so that a wavelength of λ = 1.175 Å = 0.1175 nm was obtained. The counting pulses dependent on the scattering angle were then recorded and evaluated according to the principles given above.

These measurements did not only serve to find the magnetic structure at different temperatures, they were also necessary to obtain a very good nuclear structure for later experiments.

The former were performed at the instrument D3 (ILL Grenoble) for spin-polarized neutrons. The construction of the diffractometer, electronics and additional equipment corresponds to the one of the D15 device. The generation and detection of polarized neutrons, however, require some special devices.

As a monochromator a ferromagnetic $Co_{92}Fe_8$ crystal has been used. By this, two subbeams from the primary unpolarized neutron radiation with the intensities $I^+ = (F_N + \vec{F}_M^{\perp})^2$ and $I^- = (F_N + \vec{F}_M^{\perp})^2$ are produced. By choosing a reflection with $|F_N| = |F_M^{\perp}|$—in this case it is the (200)—one of the subrays vanishes, whereas the other is linearly polarized. The wavelength amounted to $\lambda = 0.924\,\mathring{A} = 0.0924$ nm.

By excitation of the precession frequency of the neutron spins ("Meissner-Majorana-cryoflipper") with a high-frequency field, it is possible to flip these spins parallel or antiparallel to the applied magnetic field. The influence of the sample on the distribution of I^- and I^+ for a given reflection, the so-called flipping-ratio, provides the observable at these experiments.

As additional equipment we disposed of a bath cryostat with a similar temperature range as at D15 and a superconducting cryomagnet with a magnetic field of 4.62 T at maximum, which is used for the removal of antiphase domains in the samples.

3.3.4.5 Preparation of the sample (neutrons)

The single crystal specimens for our measurements have been provided by S. Hosoya, Sendai, Japan, as cylindrical pieces grown with the Czochralski method. The crystals were oriented by Laue and BPC exposures (Section 3.3.2.1) and carefully cut perpendicular to the crystallographic axes by help of an in-house constructed diamond saw. We obtained rectangular plates with the rough size of 4 × 4 × 0.5 mm^3 with the *c*-axis (SG Pnma) within the narrow edge of the crystal. The sample was aligned with the *b*-axis (assumed magnetization vector) parallel to the ω-axis of the diffractometers.

3.3.4.6 Experiments at the D15 device (ILL, Grenoble) for integrated neutrons

The measurements at D15 were performed in four steps:

(1) Establishing of the orientation matrix (cf. Section 3.3.2.3) in order to find the necessary and desired reflections.
    This step was rather easy to perform. The crystal was well pre-oriented, so that only the ω-circle had to be turned to find a strong *hk*0 reflection. Subsequently some of them were manually searched and centred around the peak maximum. The UB-matrix (cf. Section 3.3.2.3) was then optimized by several selected reflections.

(2) Confirmation of the nuclear structure by measuring a large data set at a temperature not far above the Néel point ($T_N \approx 65$ K) in the paramagnetic region.

(3) Determination of the temperature dependence of several characteristic reflections with evaluation of the exact transition temperature.

(4) Measurement of several sets of reflections (data sets) below T_N for the determination of the magnetic moments on the two crystallographic sites with size and direction.

3.3.4.6.1 The evaluation routines

Whereas the data reduction had been performed directly at the instrument yielding *hkl* listings of intensities I_{obs} and standard deviations $\sigma_{I,obs}$, the determination of the calculated intensities I_{cal} was dependent on the author's intuition.

In general, the evaluation programs work according to the method of least squares refinement. The so-called *R*-value weighted is the quantity to be minimized (correspondence parameters):

$$R_w = \sum_i \left| w_i (I_i^{obs} - I_i^{cal}) \right| / \sum_i \left| w_i I_i^{obs} \right| \tag{3.152}$$

The sum had to be taken over all reflections *i*.
The weighting scheme was provided by the connection

$$w_i = 1/\sigma_i^{I_{i,obs}} \tag{3.153}$$

by which the fact shall be expressed that very well measured reflections with low σ's should be weighted best.
Besides, we also calculated the non-weighted *R*-value

$$R = \sum_i \left| (I_i^{obs} - I_i^{cal}) \right| / \sum_i \left| I_i^{obs} \right| \tag{3.154}$$

The difference between the two *R*-values should not be too large—if the goodness of refinement is dependent on the scheme of weighting, the structure possibly might be solved badly.

As mentioned above, nuclear and magnetic intensities of integrated neutrons are mostly superimposed to give a total intensity. Their amounts are determined by the phase factors which depend on the atomic positions within the unit cell.

In the paramagnetic region above T_N the non-ordered magnetic moments will give no coherent contribution, therefore the magnetic amount at the total intensity will vanish in favour of the background—the nuclear amount will remain.

The latter one had been refined, whereas lattice constants and atomic parameters had been taken from previous X-ray data as starting values. The scale and temperature factor as well as the extinction parameters domain radius and mosaic distribution (see above, Section 3.3.4.3d) had been refined during our evaluations.

3.3.4.6.2 Data evaluation at D15

For the evaluation of the neutron measurements below T_N a special routine was created that takes the superposition of nuclear and magnetic intensities into account. Where the nuclear contribution is concerned, we applied conventional subroutines from the Cambridge Crystallography Subroutine Library (CCSL). For the magnetic contributions we set up a form factor table according to the data of Freeman and Watson (1961), the components of the magnetic moments to be refined (with respect to the crystallographic axes) and phase factors for the positions of the magnetic ions in the unit cell, separated for M1 and M2. Moreover, the q''-factor mentioned above had to be regarded, since we measured the magnetic scattering amount projected into the plane of diffraction, whereas we wanted to obtain the original F_M instead.

Another issue is the distribution of signs of the components of the magnetic moments. In the whole, we had to refer to the following atomic positions for the two sorts of magnetic ions (space group Pnma):

$$\mathrm{Me}^{2+}(M1,4a)=\begin{pmatrix}0\\0\\0\end{pmatrix}_1;\ \begin{pmatrix}0\\ \tfrac{1}{2}\\0\end{pmatrix}_2;\ \begin{pmatrix}\tfrac{1}{2}\\0\\ \tfrac{1}{2}\end{pmatrix}_3;\ \begin{pmatrix}\tfrac{1}{2}\\ \tfrac{1}{2}\\ \tfrac{1}{2}\end{pmatrix}_4 \tag{3.155}$$

The latter site has inversion symmetry $\overline{1}$

$$\mathrm{Me}^{2+}(M2,4c)=\begin{pmatrix}x\\0.25\\z\end{pmatrix}_1;\ \begin{pmatrix}\overline{x}\\0.75\\ \overline{z}\end{pmatrix}_2;\ \begin{pmatrix}x'\\0.75\\z'\end{pmatrix}_3;\ \begin{pmatrix}\overline{x}'\\0.25\\ \overline{z}'\end{pmatrix}_4 \tag{3.156}$$

with mirror symmetry m and the additional atomic coordinates $x' = 0.5-x$ and $z' = 0.5 + z$.

The original atomic coordinates x and z within the unit cell had been taken, as already mentioned, from X-ray data.

From the probable direction of previous measurements we assumed the following:

$$\vec{\mu}_{Fe}(M2,4c)=\begin{pmatrix}0\\ \mu_b\\ 0\end{pmatrix}_1;\ \begin{pmatrix}0\\ \mu_b\\ 0\end{pmatrix}_2;\ \begin{pmatrix}0\\ -\mu_b\\ 0\end{pmatrix}_3;\ \begin{pmatrix}0\\ -\mu_b\\ 0\end{pmatrix}_4 \quad (3.157)$$

The distribution of signs obeys the symmetry requirements of the magnetic group.

The calculated magnetic structure factor on M2 will then be

$$\vec{F}_M(M2,4c) = 2\mu\, f_{\text{Fe}}\begin{bmatrix}\cos 2\pi(hx+k\cdot 0.25+lz)\\ -\cos 2\pi(hx'+k\cdot 0.75+lz')\end{bmatrix}e^{-W}, \quad (3.158)$$

where μ is the magnetic moment to be refined and f_{Fe} the form factor for Fe^{2+}.

In the case of the low-symmetric (−1) M1(4a)-position, the calculation is more complicated. Below T_N, we obtain a 3D canting of the magnetic moments; therefore, 3 components along the three crystallographic axes had to be refined:

$$\vec{\mu}_{Fe}(M1,4a)=\begin{pmatrix}\mu_a\\ \mu_b\\ \mu_c\end{pmatrix}_1;\ \begin{pmatrix}-\mu_a\\ \mu_b\\ -\mu_c\end{pmatrix}_2;\ \begin{pmatrix}-\mu_a\\ -\mu_b\\ \mu_c\end{pmatrix}_3;\ \begin{pmatrix}\mu_a\\ -\mu_b\\ \mu_c\end{pmatrix}_4 \quad (3.159)$$

The calculated magnetic structure factor on M1 will then be

$$\vec{F}_M(M1,4a)=\begin{pmatrix}\mu_a\\ \mu_b\\ \mu_c\end{pmatrix} f_{Fe}\begin{bmatrix}\begin{pmatrix}1\\ 1\\ 1\end{pmatrix}+\begin{pmatrix}-1\\ 1\\ -1\end{pmatrix}\cos\pi k+\begin{pmatrix}-1\\ -1\\ 1\end{pmatrix}\cos\pi(h+l)\\ +\begin{pmatrix}1\\ -1\\ -1\end{pmatrix}\cos\pi(h+k+l)\end{bmatrix}e^{-W}, \quad (3.160)$$

By help of the factor q'' we recalculated the magnetic structure factor F_M to $F_M^{\perp}$:

$$I_{\text{cal}} = F_N^2 + q''^2\left|\vec{F}_M\right|^2 \quad (3.161)$$

We obtained for each reflection *hkl* a calculated intensity and compared it with I_{obs}; the refinement always compared the two quantities and successively improved the values for I_{cal} according to the least squares algorithm.

We also applied extinction corrections and refined scale- and temperature factor as well. From the obtained components of the magnetic moments with respect to the crystallographic axes a total moment can be calculated according to the following formula:

$$\left|\vec{\mu}\right| = \sqrt{\mu_a^2 + \mu_b^2 + \mu_c^2} \tag{3.162}$$

3.3.4.6.3 Results of the D15 neutron diffraction measurements on synthetic fayalite Fe_2SiO_4

The intensities of selected reflections, which are characteristic for spin canting, were followed as a function of temperature—the relevant points lie on Brillouin-type curves that enter the abscissa at the transition temperature antiferromagnetic/paramagnetic. Around the transition point the intensity of several reflections has been recorded in tenths of a degree, so that the Néel temperature could be determined as $T_N = 65.1(1)K$ (Fig. 3.54).

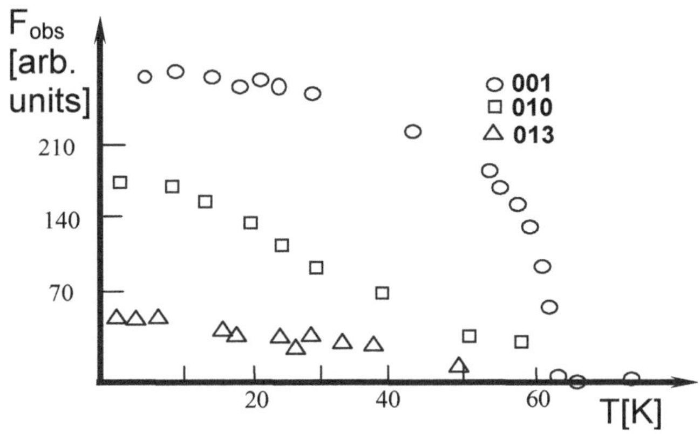

Figure 3.54 The neutron structure factors of some characteristic reflections of synthetic fayalite as a function of temperature below the Néel point.

It can be seen in Fig. 3.54 that the slope of the different curves is not uniform. This indicates a non-continuous decrease of the canting on the M1 position. The recording of large data sets, however, at 4 K, 23 K, 35 K and 50 K gave no hint to the assumed peculiarity at $T \approx 23$ K, which was found in previous magnetometric measurements (Ballet et al., 1989).

Table 3.2 shows the refined magnetic moments for the two non-equivalent crystallographic sites M1 and M2, together with the direction cosines of the components of the magnetic moments with respect to the crystallographic axes. The *R*-values point to a quite good quality of refinement.

Table 3.2 Magnetic moments and normalized directions of moments for four temperatures in the antiferromagnetic state of synthetic fayalite on the two iron sites M1 and M2 below T_N = 65 K derived from integrated neutron diffraction measurements[15]

T [K]	10	23	35	50
μM2 [μ_B]	4.00(5)	3.72(17)	3.83(4)	3.38(15)
μM1 [μ_B]	3.88(10)	2.92(13)	2.60(14)	1.67(17)
cos (μ,a)	0.31(2)	0.24(4)	0.28(6)	0.10(6)
cos (μ,*b*)	0.77(1)	0.85(3)	0.89(1)	0.94(9)
cos (μ,*c*)	0.56(2)	0.48(3)	0.38(7)	0.33(4)
R [%]	3.6	6.2	3.0	5.5
R_w [%]	3.3	6.4	2.7	6.4

As it can be seen, there is a collinear antiferromagnetic ordering on the M2 position along the *b*-axis (space group Pnma), whereas on M1 we observe a canting of the magnetic moments dependent on temperature in the range of 4.2 K $\leq T \leq$ 50 K. Above T = 50 K, both total magnetic moments on M1 and M2 should be aligned parallel to **b**. The deviations of the total magnetic moment from the expected value of 4 μ_B are attributed to a strong contribution of orbital moments.

3.3.4.7 Experiments at the D3 for spin-polarized neutrons

It should be mentioned in advance that we do not dispose of polarized neutron data on synthetic fayalite—the crystal sample of my predecessor at Grenoble revealed a severe 50:50 distribution of antiphase domains (thus flipping ratios close to 1) so that his data could not be evaluated.

[15]Table according to Müller (1985).

However, we plan the repetition of this experiment on another sample in near future. However, since a polarized neutron experiment for the support and comparison of nanoscopic results is extremely reasonable, the reader should at least learn the basic principles of a polarized neutron refinement. Therefore, here we report our analogue example of synthetic tephroite Mn_2SiO_4.

Also in this case, a fixed measuring cycle was prescribed:

(1) Determination of the orientation matrix in order to find the necessary reflections. Since the equipment of the diffractometer D3 is similar to that of D15, we aligned the crystal with its *a*-axis parallel to the vertical ω-axis and, at the same time, parallel to the polarization vector of the incident neutrons and the axis of the applied magnetic field. The UB-Matrix was subsequently established. By manual peak search and centring of the chosen reflections, it was quickly achieved to optimize the UB-matrix.
(2) The external magnetic field of 4.62 T parallel to the *a*-axis was applied in order to remove possible antiphase domains.
(3) Measurement of the flipping ratio (fr) for a big number of reflections at a fixed temperature.

3.3.4.7.1 The evaluation routine

The basic version of this program was created by P. J. Brown (ILL, Grenoble) and was intended to perform a preliminary evaluation of the measured fr's for collinear antiferromagnetic structures. From the experimental flipping ratios the γ-values could be calculated—after determination of the crystallographic structure with well-established nuclear structure factors F_N the magnetic structure factors F_M could be derived by help of the γ-values. As a result, one obtains a *hkl* listing with F_M^{obs} and σF_M^{obs} in units μ_B; F_M is not written here as a vector, since only the magnetic moment direction parallel to the spin alignment of the incident neutrons can be determined.

Within the conventional program it was also possible to calculate a magnetic structure factor $F_{M,cal}$, which, however, proved to be too simple to meet the experimental data. For example, we observed a symmetry break yielding a difference in allegedly equivalent fr's and F_M's for *hkl* and *hk*–l reflections. The easy antiferromagnetic model provided by the basic refinement program could not reproduce this asymmetry.

Therefore, an almost new magnetic structure refinement program had to be written, which shall be presented in the following:

In principle, this routine consists of two parts:

(1) Calculation of the F_M^{obs} from the measured fr's. From them, the factor γ is derived according to the following equation:

$$\gamma_{1,2} = \frac{P(1+RE)}{1-R} \pm \sqrt{\left(\frac{P(1+RE)}{1-R}\right)^2 - 1/q''^2}, \tag{3.163}$$

where P is the polarization grade, E the cryoflipper efficiency and R the flipping ratio.

Subsequently, the nuclear structure factor is determined as described above. It should be mentioned again that the crystallographic structure has to be determined accurately before a measurement with polarized neutrons takes place (e.g., by application of integrated neutrons), since via F_N the observed magnetic structure factor is derived by

$$F_M^{obs} = \gamma F_N. \tag{3.164}$$

This means that poor F_N's will produce poor F_M's. The parameters P and E in the above equation are device constants, R is the measured (observed) fr. Since F_M is permanently $<F_N$, we take of both possible γ-values the one which is lower than unity.

(2) Determination of the calculated magnetic structure factors F_M^{cal}:

As a basic structure, we take the spin arrangement analogous to that above for Fe_2SiO_4 (cf. Section 3.3.4.6.3). The expression for F_M^{cal} on M1, however, is simplified, since we only "see" the collinear contribution parallel to **a** with polarized neutrons—therefore we only have to calculate the component of F_M^{cal} parallel to **a**. Concerning the phase factors, the least-square-subroutine and the correspondence factors R and R_w as well as the output of the magnetic moments, everything is analogous as above.

During the refinement it turned out that an additional ferromagnetic moment had to be introduced. This was regarded by an additional ferromagnetic contribution in the total structure factor; the former was furnished with the same phase factors as the antiferromagnetic contribution, however, with two ferromagnetic moments per site.

A further option was the introduction of a correction for domain structures and a possible magnetic moment on superexchange oxygens by the formula of Bonnet (1979).

3.3.4.7.2 The Fourier program

This routine stems from the CCSL (ILL Grenoble) and performs a Fourier back transformation resp. inversion of given magnetic structure factors by the following equation:

$$\rho_S(\vec{r}) = \frac{1}{V_0} \sum_{\vec{g}} e^{-\vec{g}\vec{r}} \vec{F}_M(\vec{g}) \tag{3.165}$$

This leads to a position-dependent distribution of the magnetic moment in the unit cell, the so-called spin density.

Since this density is three-dimensional (exactly speaking: four-dimensional, because the density is dependent on three coordinates x, y, z), in former times suitable sections or projection planes had to be chosen in order to draw the well-known 2D contour maps.

The position of these planes can be provided by rather simple parameters within the program. As a second step the points of same spin density values are detected and connected by contour lines. One then gets a contour map with (in the isotropic case) concentric circles around the positions of the magnetic ions. The decrease of the spin density around the centres of the charge distribution—visible by the distance of these lines to one another—is determined by the form factor involved.

A further option is the possibility to draw a difference density: Besides the listing of F_M^{obs} another table of F_M^{cal} has to be put in. Instead of only experimental magnetic structure factors the difference of observed anisotropic and calculated (isotropic) structure factors is used in order to obtain a "non-compensated" spin density arising from all spin density distributions which are not theoretically foreseen. By this, completely unexpected details can be detected under the condition to dispose of a sufficiently large data set and a critical review of the obtained contour map.

3.3.4.7.3 Remarks concerning the data evaluation at D3

The performance of the above programs was not straightforward as the previous chapters might infer. During the evaluation new aspects, sometimes surprisingly, arose, which led to changes in the programming with new refinements and insights—this "iterative" procedure, however, led in the end to a quite self-consisting image of the derived properties and minimum *R*-values.

In the beginning of the procedure, the experimental features seemed to be inexplicable and it was understood why there are comparably few measurements of polarized neutrons on antiferromagnetic compounds, although this magnetic ordering is widespread in nature.

Since the experiment was performed at $T = 20$ K, the almost collinear antiferromagnetic structure model obtained with integrated neutrons should have been successful. However, this led to *R*-values of 30% and more. The observation of asymmetric reflection groups with $F_M^{obs}(hkl) \neq F_M^{obs}(hk{-}l)$, however, let us suppose that the rather simple magnetic model could not be applied in this case—obviously the magnetic group symmetry had been severely lowered.

Closer inspection showed that this was caused by the external magnetic field that was applied parallel to **a** in order to remove antiphase domains. According to the theory of antiferromagnetism (cf. Lottermoser, 2015) a magnetic field parallel to the direction of the magnetic moment enhances the latter at the expense of the antiparallel oriented one—the thus induced ferromagnetic moment well corresponds with the observed reduction of symmetry. The additional ferromagnetic moment introduced in the refinement is fully compatible with the symmetry requirements.

With these changes a pronounced improvement of the R_w-value down to 9% was achieved. The observation of structure factors $F_M^{obs}(0kl)$, which should be forbidden in the pure antiferromagnetic case, is a direct hint to the ferromagnetic contribution mentioned above.

However, we still obtained antiferromagnetic moments which were much lower than expected. We attributed this problem to still existing antiphase domains which should not have been fully removed.

Indeed, the introduction of these domains led to success: The data set was refined under different ratios of antiphase domains till we attained values of the magnetic moments that were reasonable in view of our integrated neutron results.

This is, however, not self-evident: the ratio $|\mu(\mathrm{M1})|/|\mu(\mathrm{M2})|$ could have been shifted by the domain refinement, so that no domain ratio would have yielded a reasonable value for the magnetic moments. Moreover, magnetometric data by Lottermoser and Fuess (1992) showed that these domains were no simple fitting quantity, but existed indeed.

As a last step of the refinement we introduced an additional magnetic moment on oxygen positions. Up to then, it could not be explained from where the large difference of the size of the two magnetic moments on M1 and M2 arose. This moment transfer due to superexchange coupling was predicted according to solid-state theory (cf. Lottermoser, 2015) and it was also found experimentally by the study of Bonnet et al. (1979). Interestingly, this possibility only led to success in the refinement, if not every oxygen was taken as a bearer of moment, but only those, which have been predicted as candidates for superexchange according to the previous study of Fujino et al. (1981).

3.3.4.7.4 Remarks concerning the Fourier program

As is well known from the theory of Fourier inversions, the absence of certain Fourier components (here, missing structure factors of certain *hkl*'s) would lead to series termination errors. For example, in the common case of Fourier synthesis of a rectangle, the latter errors present themselves as "Gibb's ears". Also in our case, the Fourier inversion with an incomplete data set of structure factors may cause strong series termination errors.

This problem is widely underestimated in literature: It leads not only to distortions of the spherical symmetrical distributions of spin density but also to completely artificial "peaks" which even do not exist in reality. We might call them "Fourier ghosts". These have been found during our first trials with the conventional Fourier inversion program and could be assigned to the comparably high-symmetrical M2 position. Later, during our studies concerning our nanoscope, we independently detected such Fourier ghosts, but managed to "buster" them (cf. Section 5.4). Obviously, the number of our structure factors was too low to avoid such undesired effects.

In this situation, we applied a trick: Since our magnetic structure model was sufficiently good and R_w low, we filled up the experimental data set of 119 F_M^{obs} with several thousand F_M^{cal}; however, the contour maps will then only show the principal features of the refinement, smaller details should be overwhelmed by the multitude of F_M^{cal}. A classical difference Fourier synthesis would not lead to success either, because the difference F_M^{cal}–F_M^{obs} would then again lead to small deviations of 119 reflections—these differences again would be subject to series termination errors.

Hence, we must not apply a Fourier inversion to the above difference, but we have to perform one Fourier inversion of the complete data set of F_M^{cal} and one of the "refilled" data set of F_M^{cal} -($F_M^{cal}(hkl_{obs})$ + F_M^{obs})! In this "refilled" data set all these F_M^{cal} were avoided, for which F_M^{obs} existed—the structure factors were then linearly overemphasized by a huge scale factor. Subsequently, the two spin density maps were compared with each other. Thus spin densities arose, that were due to the few F_M^{obs}, but—due to the large number of structure factors being processed, the maps did not suffer from series termination errors.

To summarize, we might state that Fourier spin densities and difference densities only lead to reliable results, if a very large number of experimental structure factors is available. If this is not the case, missing strong F_M^{obs} may lead to Fourier ghosts as a cause for erroneous interpretations.

In case of a very good structure model (low *R*-values), however, it is possible to apply "refilled" data sets according to the prescription given above. Analogously, the above conclusions are also valid for electron densities from X-ray structure data.

3.3.4.7.5 Results of the polarized neutron refinement

The multitude of F_M^{obs} and F_M^{cal} could be divided into groups which were characterized by different magnetic contributions: one antiferromagnetic and two ferromagnetic ones per site (the latter due to the applied magnetic field). The domain refinement yielded a ratio of antiphase domains of 92:8; the sample may be considered as almost containing a single domain, which could be expected in view of the strong magnetic field and the low anisotropy. The corresponding magnetic moments are given in Table 3.3.

Table 3.3 Different contributions to magnetic moments of synthetic tephroite Mn_2SiO_4 in the antiferromagnetic region on M1 and M2 at T = 20 K derived by polarized neutron diffraction measurements

T = 20 K			
Antiferromagnetic moment [μ_B]			
μ(M1)		μ(M2)	
2.56(2)		4.33(2)	
Ferromagnetic moment [μ_B]			
μF + (M1)	μF − (M1)	μF + (M2)	μF − (M2)
0.38(2)	0.16(2)	0.14(2)	0.10(1)
Magnetic moment on oxygen positions [μ_B]			
μ(O1)	μ(O2)	μ(O3)	
—	0.48(8)	0.26(9)	
R [%]	7.1		
R_w [%]	7.5		

The antiferromagnetic moments correspond well with those found by the integrated neutrons. The ferromagnetic contributions are larger on M1, which fits to the lower local symmetry (−1) and the residual canting along **b** and **c**. These findings are in good agreement with our expectations.

However, what is about the special feature of polarized neutrons, i.e. to establish an experimental form factor μ_f dependent on sineθ/λ? In order to achieve this, the F_M^{obs} had to be divided by the relevant phase factor. In case that the experimental F_M^{obs} can be attributed unequivocally to the existing crystallographic sites, there will be no difficulty. However, if the F_M^{obs} may consist of mixtures of components with different phase factors, the problem cannot be solved. On the other hand, the F_M^{obs} are so well refined that the F_M^{cal} well represent the relevant experimental data. In Fig. 3.55 we show these form factor curves under domain refinement for the two crystallographic sites M1 and M2:

We may observe here the form factor decrease for the two different magnetic moments on M1 and M2 dependent on sineθ/λ. These curves will evidently contain no information about covalency effects. However, what is about the refinement of the magnetic moment on oxygen positions?

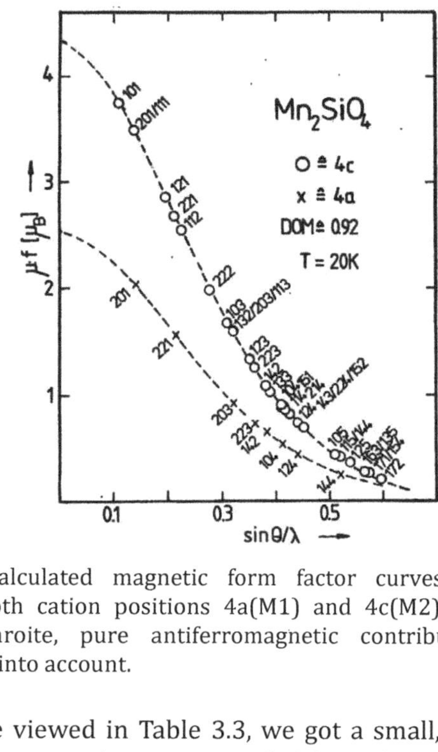

Figure 3.55 Calculated magnetic form factor curves dependent on sineθ/λ for both cation positions 4a(M1) and 4c(M2), respectively, in synthetic tephroite, pure antiferromagnetic contribution, antiphase domains taken into account.

As can be viewed in Table 3.3, we got a small, but significant magnetic moment on those oxygens (O(2) and less, on O(3)), which were supposed to transfer superexchange coupling. These are anions O^{2-} which form large angles >>90° with the positions of the metal cations on M1 and M2, respectively—the *R*-values were additionally improved by this measure.

Are there other hints for these moments on oxygen positions? We claim here the contour maps provided by the Fourier inversion of the different data sets of structure factors (as mentioned above).

We can rule out Fourier ghosts since this effect is not probable with respect to the huge number of structure factors (4630) being used. In order to make as few premises as possible, we established the relevant two contour maps within the space group with lowest possible symmetry, P-1—the relevant information is solely due to the magnetic structure factors $F_{\mathrm{M}}^{\mathrm{obs}}$ themselves (Figs. 3.56 and 3.57).

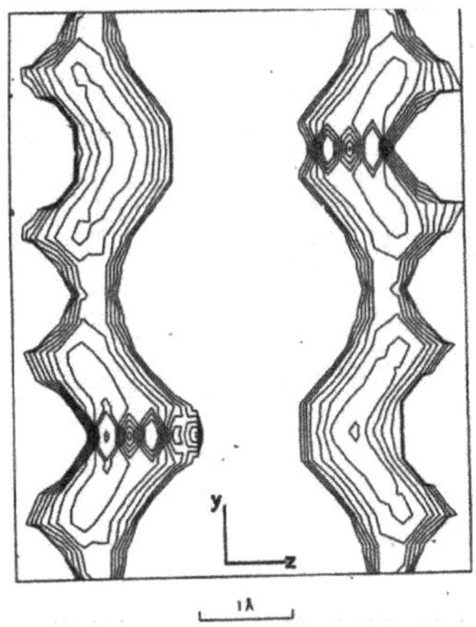

Figure 3.56 Magnetic moments' density in synth tephroite for the O(2) position. Cut through the (011) plane (SG Pnma) for x = 0.45, based on pure calculated structure factors. Distance of contour lines: 0.001 $\mu_B/\text{Å}^3$, starting with 0.01 $\mu_B/\text{Å}^3$, T = 20 K.

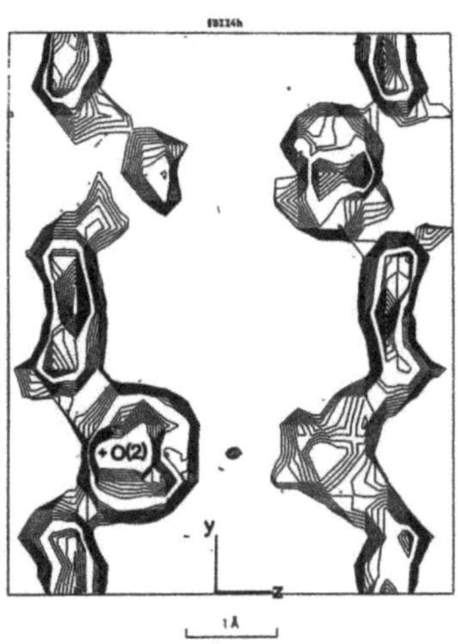

Figure 3.57 Magnetic moments' density in synth. tephroite for the O(2) position as described in the figure on the left. Observed and calculated structure factors regarded according to the method in the text. Contour lines as in the above figure, T = 20 K.

Comparing the maps with each other, we indeed notice a non-vanishing magnetic moment on O(2). This peak is overemphasized by the large scale factor, but nevertheless, the contours of the map based on pure $F_{\mathrm{M}}^{\mathrm{cal}}$ are pronouncedly different to the map of the "refilled" data set—the difference arises from the low amount of $F_{\mathrm{M}}^{\mathrm{obs}}$ (there is no other possibility)—therefore we believe in the correctness of our refinement which will be later confirmed by our nanoscope images (Section 5.4).

Chapter 4

The Extension of Pillar 3: The DEN Method

4.1 The Principal Idea

For my thesis work in the early 1980s, I already was working on several synthetic olivines, fayalite α-Fe_2SiO_4, tephroite α-Mn_2SiO_4 and cobalt orthosilicate α-Co_2SiO_4, mainly with the comparably new and exciting method of neutron diffractometry. In order to align and cut the kindly provided single crystal specimen from Dr. Hosoya, Japan, and his group, I had to learn basic crystallography and X-ray diffractometry—not so easy for a just-finished physicist with diploma certificate whose ink had not yet dried.

The crystallographic axes of the crystals had to be found by X-ray methods like Buerger precession (I dedicated Section 3.3.2.1 to the description of the camera and its handling), since the suitable neutron diffractometers at Institute Laue Langevin (ILL), Grenoble, were in fact two-circle instruments. Hence, the crystals had to be oriented—fortunately they all were orthorhombic—and cut with a diamond saw into the "normal shape" of a neutron-suited sample, i.e. a flat plate (to minimize absorption effects) of about 0.5 mm thickness and about 4 × 4 (or more) mm^2 within the plane. The edges should be cut possibly along two of the three crystallographic axes in order to easily retrieve the axes later on the neutron diffractometer.

The Difference Electron Nanoscope: Methods and Applications
Werner Lottermoser

ISBN 978-981-4774-01-7 (Hardcover), 978-1-315-19664-0 (eBook)
www.panstanford.com

The first day at ILL (our most accessible neutron facility) always was pure stress for the external user (like me): Was the crystal properly aligned or not? In the second case a good deal of the valuable measuring time would have to be "wasted" for the peak search; in this case it could happen that a further proposal was "not eagerly" approved (to say the least of it)—a nightmare for a thesis student who was dependent on these experiments in—minimum—half of a year timely distances. Fortunately, the "UB-matrix" for my crystals had been found almost instantaneously, and the collection of the data sets could then be started—there was always great relief. At this point I should mention my Senior Scientist at the D15 and D3, my very dear P. Jane Brown, who mainly performed the experiment and never lost patience with me "greenhorn". I am also indebted to my old friend and predecessor Rolf Müller, who taught me everything what I had to know about neutron diffraction and last, but not least, to my dear supervisor, famous Hartmut Fuess from Frankfurt/Darmstadt.

So I have learned a lot about diffraction and—in another part of my thesis work—about Mössbauer spectroscopy, since Fe_2SiO_4 as an iron bearing compound is sensitive to ^{57}Fe MBS. These measurements were performed at Marburg, Germany, and also evaluated at Grenoble, because at Marburg such a powerful program to refine a two-subspectra hyperfine pattern with 16 parameters was not available in those days. I also learned a lot about spectroscopy (courtesy of Georg Amthauer there) and this "magic" basic quantity of MBS, the electric field gradient.

Perhaps due to my previous intense occupation with diffractometry and Fourier representation of electron densities it came to me, whether it might be possible to calculate an efg from electron densities—in this case 3d electron densities, since Fe is a 3d metal.

This idea was discussed with Stefan Hafner, the Leader of the Mössbauer Lab at Marburg in those days, and he kindly invited me to present the idea at a small colloquium that he was organizing at Marburg. I was prettily excited during my short talk, since there were experts like Helmut Rager (esr, experimental efg's) and Karlheinz Schwarz from Vienna (famous theoreticist) in the audience; but there was a lot of approval and no severe objections, at least according to my impression.

During my time of habilitation at Salzburg University, I mostly developed the experimental part of the idea—the evaluation of the efg from single crystal Mössbauer spectroscopy (SCMBS), but also did some modest trials to calculate an efg: Thomas Armbruster and Martin Kunz from Bern, Switzerland, kindly provided electron density maps from their excellent X-ray data refinements of different samples. In those days, it was rather arduous to derive an efg from these contour maps—I copied them to transparencies, put them on a computer screen and tried to trace them by mouse and cursor—the beginning of digitizing of a map. The results at least gave the main component and the size of the efg rather reasonably. Later on, after habilitation, I applied for a project at FWF (Austrian Fund for the Scientific Research) in order to make things properly and to develop a program that should derive an efg from three-dimensional 3d electron densities stemming from diffraction data. The obtained "semi-empirical" efg should then be compared with the experimental one from SCMBS and DFT calculations. This was quite a heavy enterprise, but—fortunately—I had the best imaginable people "on board": Konrad Steiner, mathematician (and biologist) and best thesis student ever; Gerhard Scharfetter, computer scientist; Hannes Leeb (mathematician and Fourier specialist) and last but not least, Michael Grodzicki (theorist and DFT expert) and Armin Kirfel, crystallographer and provider of the excellent fayalite synchrotron diffraction data.

After the successful finishing of the project, we all knew the difference between an idea and the practical realization of it: It comprised several years' work and a computer code of around 3000 lines.

4.2 The Hardware Components

The decision which machine to afford in order to run the future programs, certainly being very sophisticated, was not easy and, in the beginning, controversially discussed in the group. Workstation? Central computer with terminal? PC? The hardware had to meet several conditions, which partly were based on previous experiences:

(1) The computer should dispose of an "open" system and architecture, i.e. no software-inherent limitations of memory space.
(2) The machine should also physically allow to make later extensions of memory space (i.e. possibility in the computer housing for additional hard discs, RAM memory, a.s.o.).
(3) The hardware should, by no means, develop an unpredictable, independent existence like frequent system crashes without obvious reasons, blue screens, freezing, a.s.o.—unnecessary obstacles that would delay the programmer's work and drive the group to madness—we have had our experiences before and I do not explicate this point further from judicial reasons.
(4) The decision which software frame to take, was made earlier—hence the new machine should be compatible with this software which, vice versa, should offer a documented version for the type of computer.
(5) The machine's system software should be easy to learn, optionally with multiple access under other systems for professionals, like UNIX or LINUX.
(6) The computer (or access to central computer) should be "affordable" in the frame of the project—this ruled out high-end machines or big workstations.
(7) Last but not least: It should be fast enough to handle large data sets in a reasonable time and also, to display large images possibly real-time on the screen—this includes a sufficiently powerful graphics card.

In particular, taking the last point into account, there remained only one realistic candidate: The Power Mac G5 (or Power Macintosh) by Apple Computer Inc. with double processor—the fastest and, in my opinion, best "private" or semi-professional computer on the market at that time. According to my knowledge, there was for a long time before our purchase a ban on the export of the G5 machine by the U.S. authorities—the G5 was so powerful and, at the same time, compact and comparably inexpensive that the U.S. government worried about a transfer to "rogue states", where the machine could have been used for missile guidance purposes. But fortunately for us, shortly before the launch of our project, this confinement had been abandoned.

But let us now consider the computer itself in more detail:

When introduced in June 2003, the machine was described as the fastest personal computer ever built. Even the case of it was a novelty: It was made from anodized aluminium with four integrated handles at the four narrow edges (easy for transportation) and massive aluminium covers at the side of the tower. Front and back were perforated by a multitude of small holes enabling large air feeding; from this feature the machine received its nickname "cheese grater". Behind an unimpressive flap at the front side, the G5 contains a Superdrive optical disc drive; a 250 GByte hard disc is provided inside the case as standard equipment. Figures 4.1 and 4.2 show the outer and inner view of the G5, respectively[1]:

Figure 4.1 Apple Macintosh G5, outer view with power switch, USB ports and Superdrive optical disc drive behind the upper aluminium flap.

[1]I, the copyright holder of this work, hereby publish it under the following license: *(Bernie Kohl)*. This file is made available under the Creative Commons CC0 1.0 Universal Public *Domain Dedication*. The person who associated a work with this deed has dedicated the work to the public domain by waiving all of his or her rights to the work worldwide under copyright law, including all related and neighboring rights, to the extent allowed by law. You can copy, modify, distribute and perform the work, even for commercial purposes all without asking permission.

Figure 4.2 Apple Macintosh G5, inner view, as described in the text. A similar computer is used by the author and co-workers.

The machine is based upon a dual-core POWER4 microprocessor from IBM—the original processor had 50 million transistors and was sold in different versions, lastly even with a quad-core.

Ours, however, contained two Power PC 970 FX microprocessors (Fig. 4.3[2]) with 64 bit addressing, a novelty for a PC at that time as well.

Figure 4.3 PP970 processor as described in the text. Permission for reproduction given under the terms mentioned in footnotes 1 and 2.

[2]Permission is granted to copy, distribute and/or modify this document under the terms of the GNU Free Documentation License, Version 1.2 or any later version published by the Free Software Foundation; with no Invariant Sections, no Front-Cover Texts, and no Back-Cover Texts. A copy of the license is included in the section entitled *GNU Free Documentation License*. Subject to disclaimers.

Each processor has two 32-bit pathways to and from the processor, respectively, which enables a data rate of up to 20 GByte/s (very high velocity, one of our requirements!).

The PP 970 processor can address 2^{42} bytes of physical memory and 2^{64} bytes of virtual memory. By its 64-bit addressing it can also administrate up to 8 GB Dual-Channel DDR PC3200 SDRAM space. Fast graphics are provided by a NVIDIA GeForce FX5200 ultra card with 256 MB DDR RAM. The hard drive runs on 7200 rpm (Serial ATA = SATA standard), the optical drive can read and write DVD's and CD's with up to 32× speed. The G5 contains several expansion slots (as desired by us) and a lot of peripherals, among them 3 × USB 2.0, 2 × Firewire 400 and 1 × Firewire 800—these are Apple standards and rather important, since it might have been necessary to use external memory for input and output of data—the fastest computer may severely be "braked down" by slow peripherals. This is not the case with the G5.

The system used is MacOSX 10.4 "Tiger"—a very comfortable operation system (OS) with GUI (graphical user interface) and possibility to use a virtual UNIX terminal (implemented in the GUI), also one of our requirements. In order to have an optimal connection between the display and the computer, we chose a 20″ Apple wide screen as the visual unit.

The commercial software frame (IDL, see Section 4.3.1) could be implemented without problems, so that, after several years of operation, we can say, that we never regretted our initial decision.

4.3 Description of the Software

The choice of the suitable software program or frame meeting our aims was even more difficult than the selection of the hardware—in fact, this most important step of finding a program system tailored for our purposes was performed in the very beginning of our project. At this point, I should summarize which requirements should have been fulfilled by the desired software:

(1) The general aim was to perform a Fourier inversion of a large number of experimental structure factors together with their phase angles in order to obtain a three-dimensional

screen plot of the resulting electron or difference electron densities (DED's) according to the equation

$$\rho(x,y,z) = \sum_h \sum_k \sum_l A_{hkl}\, e^{-2\pi i(hx+ky+lz)} \tag{4.1}$$

where A is the structure amplitude $|A_{hkl}| = |F_{hkl}|/V_0$ and x, y, z the fractional coordinates with respect to the unit cell.

The "difference Fourier" is calculated with the rather simple equation:

$$\rho_{\text{aspherical}} = \rho(F_{\text{obs}} - F_{\text{cal,spherical}}) \tag{4.2}$$

The rough idea was to establish a semi-empirical (or semi-quantitative) efg from this aspherical density, since it should contain the 3d electron contributions.

(2) The desired software should display the electron densities or DED's as a function of the three coordinates x, y, z, possibly together with the site of the nuclei for the viewer's better guidance within the unit cell. In this sense, the density is in fact four-dimensional.

(3) An integration of chosen difference electron densities over the relevant space sector should be performed in order to derive an efg tensor (size and direction of the main axes) according to a point charge formalism.

(4) If possible, the efg tensor should be displayed in an ostensive form within the same simultaneously displayed unit cell.

(5) The desired software should run real-time (or almost real-time) and ought to be manageable not only by computer freaks.

Fortunately, my thesis student had a lot of connections in mathematics, computer sciences and other relevant fields: The best choice for the software frame which met most of the above requirements was a surface which stems from the field of computer tomography. A colleague from the geology department used this software successfully in order to display rocks and their constituents three-dimensionally.

It is based on an own programming language (IDL = interactive data language) and has the big advantage to provide tailored solutions. Perhaps there were already complete solutions on the market at that time, but these were mostly unattainable in price or much too specified or, on the contrary, too simple.

The IDL package version 5.6 was sold by RSI (Research Systems Inc. (2002))—IDL is a registered trademark; details may be taken from the user manual "Getting Started with IDL".

In the Introduction section of this manual, all important features are mentioned: IDL is claimed to be the "ideal software for data analysis, visualization and cross-platform application development". It is described as a "fourth generation language", "more compact than C or FORTRAN" enabling a giant spare of time compared to these much older languages. It implies a lot of mathematical analysis (already predefined) and physical display techniques providing "incredible flexibility". We can confirm everything of this enthusiastic description from own experiences. For readers who might be interested in this software frame, please find more details in the following section.

4.3.1 The Commercial Software Frame IDL (Interactive Data Language)

IDL is a very flexible and mighty tool for all kinds of data processing and displaying:

(1) It is possible to apply IDL interactively using functions and procedures in order to create data.
(2) There is support for many scientific data formats and image standards to read and write data.
(3) It contains signal processing procedures to process 1D signals using filtering and transform options, to make 2D contour or surface plotting and 3D-volume visualization using shaded or constant density (iso-) surfaces.
(4) It is possible to do image processing for noise filtering or highlighting special data characteristics—the latter was a very nice tool in our case to select different electron densities for the calculation of the efg.
(5) It can transform "irregularly sampled" data to a regular grid.
(6) Animation possibilities: Dynamical visualization of data in order to create an array of images and to show a "movie" —I did this for a talk in order to let the audience get a real 3D-impression of electron densities by moving the virtual

camera around the centre of the unit cell. Normally, the intrinsic 3D visualization of IDL is quite fine, but a flat screen remains 2D, despite 3D-images being shown on it; it is often impossible for the viewer to decide, which density peak is in the foreground and which is in the background. The animation gives at once an impression about the position of the DED's in space (closer ones move faster than more distant ones!).

(7) The last point mentioned here summarizes some features of IDL simultaneously:

With the help of a GUI builder, it is possible to create well-arranged, tailored windows for the non-computer freak. The programmer has another GUI (IDLDE: IDL Development Environment) at hand with a complete set of program-control statements in order to compile and execute commands immediately (including editing and debugging)—in this sense IDLDE merely provides an interpreter instead of a compiler.

We were widely using this mighty tool for our application: a program to display DED's and to graphically integrate them in order to calculate and display efg's in the unit cell.

4.3.2 The Preparing Crystallographic Routine EVOX

Before starting our main project, we had to solve a problem that is mainly due to the special properties of crystallographic systems. I recall that we dispose of seven crystallographic systems with different shapes of the unit cell: cubic, tetragonal, orthorhombic, trigonal, hexagonal, monoclinic and triclinic. With the first system, cubic, we would not get much trouble to display a "point" or "position" in three dimensions; this is commonly denoted as "voxel", analogous to the well-known "pixel", its two-dimensional variety.

A voxel in the cubic system is just a tiny cube; it depends on the desired resolution and abilities of the underlying program, how "tiny" this cube is. In any case, it has equal edges and orthogonal angles.

But how to handle a voxel in a tetragonal or rhombic system, where one or all three edges have no equal sizes? Or, even more "awfully" to imagine, what about the last four systems mentioned above, where also the angles are not orthogonal? Such "voxels", at first intuition, would resemble the underlying oblique unit cells.

The objection, however, is: Voxels in computer programming are never oblique or unequally sized. A quadratic pixel on a 16:9 widescreen remains a quadratic, and not rectangular pixel.

In this situation we had to find an algorithm which adapts the cubic voxels to the different possible shapes of the unit cells. For example, in the case of synthetic fayalite, we consider an orthorhombic unit cell (fortunately, no oblique cell!) and virtually fill it up with cubic voxels. However, the unit, in which the electron densities are displayed, is not the unit cell, but a predefined cube with an equal edge length of several Ångström units. This cube contains in its centre the ion, around which the different hyperareas of electron densities or DED's are produced. Normally, these are sites of great importance concerning the calculation of the efg's, e.g. M1 or M2 in synthetic fayalite. The cube may be shifted to other centre positions in the unit cell, so that always a suitable reference point may be chosen.

Now, there was another issue:

It should be mentioned that by enhancing the resolution of the calculation and the display by the factor 10 along the different directions, and thus reducing the voxel size by the same factor, the complexity of the calculation and the amount of necessary memory space has to be enhanced by the factor 1000. Therefore I already ask for understanding when the DED images presented here (Section 5.4) are not so well-resolved as sometimes shown in airbrush images. The limited memory space at the times of the generation of the program is the reason for it.

Another task to solve in the same routine was the calculation of the aspherical experimental electron density from structure factors F_{obs}, the spherical density from the F_{cal} and the difference electron density DED from the Fourier inversion of F_{obs}-F_{cal}, altogether taking the phases into account.

For that, long listings of reflections with *hkl*, F_{obs}, F_{cal}, F_{obs}-F_{cal} and phase had to be processed. Normally, the Fourier inversion of these structure factors would have consumed a lot of processor time—hence, we programmed a specially adapted inverse Fast Fourier Transform (IFFT) in order to considerably reduce the time need. Contrary to the direct calculation, a FFT uses intermediate results by which repeated arithmetic operations can be avoided.

the most widespread method stems from Cooley and Tukey (1965); a first version, however, of a similar algorithm was presented by Carl Friedrich Gauß (1805) for the trajectory calculation of several asteroids.

The input file of the EVOX program is a bit tricky: It consists of a string of parameters given in the UNIX shell—EVOX is a program written in C.

The first parameter in the string is the name of the input data file (for the contents see next subchapter below). Further, one has to know which reference point in the unit cell to take. For that, it is necessary to know the atomic positions within the cell from a previous structure refinement.

For that purpose, one may take a suitable structure display program, e.g. Diamonds or else. We took CaRIne, a very flexible and mighty system from France (Boudias and Monceau, 1989–1998): It allows to choose the given space group and then, by help of a virtual Mendeleev table, to put in the different types of ions and their positions (atomic coordinates) within the cell, eventually the occupation factor of the relevant site. From the symmetry operators of the space group, the program automatically generates all equivalent positions of the different ions. The size and angles of the unit cell are also requested, so that CaRIne can make a screen plot of the cell and its contents (displayed as shaded balls with different colours). The cell can now be easily viewed from all kinds of perspectives. Also, there are a lot of tools available: It is possible to search for and mark special building units like tetrahedra or octahedra (together with their surfaces); in several other windows, one may plot a theoretical powder diffractogram, stereograms, reciprocal lattice planes, a.s.o. The last feature is in particular interesting since it is possible to compare these images (e.g. a plane *hk*0) with the experimental exposures of a BPC (see above, Section 3.3.2.1), providing even the relative intensities, from which the conditions of limiting possible reflections (vice versa: cancellation rules) might be derived.

However, turning back to our main topic, it is possible to easily find the coordinates of the desired reference point just by editing the fractional coordinates of the relevant ion or point in space and to transform it to absolute coordinates by multiplying the fractional ones by the lattice constants in the three orthogonal directions.

From this absolute centre point, we—in case that we might wish to consider a 6 Å cube around the centre—have to subtract 3 Å from the x_{centre} to get x_{start} and to add 3 Å to x_{centre} to get x_{end}; these are the two coordinates along the x (or **a**) direction. It follows a number, which gives the resolution of the desired grid. For example, the value 200 means that we produce 6 Å/200 different grid points, starting with x_{start}; analogously, the parameters for the y and z directions are put in. Then we have to give the lattice constants in the order of the space group and the three angles (in our case: 90°). The last parameter is the name of the output file which might be binary or in ASCII-code.

By starting EVOX this lastly mentioned file is produced which is the later input file for our main DED routine.

4.3.3 The Input of the Experimental and Calculated Structure Factors

In fact, these data are provided by the first parameter (name of the input data file) mentioned above in the previous section. It contains the integer values of the reflections h, k, l and either the associated observed structure factors F_{obs} or F_{cal} (for a separate display of the nuclear positions or "core file") or F_{obs}-F_{cal} in order to generate a "difference Fourier" i.e. the aspherical electron density as the source of the efg.

A last column provides the associate phase angle for each reflection: Hence, dependent on the desired outcome, the file is created from the chosen diffractometer data and/or a spherical refinement (F_{cal}).

4.3.4 The Main Program DEDLOT and Its Mode of Operation

The program DEDLOT (Difference Electron Density/Lottermoser) is the result of several years' intensive discussions, IDL programming and test runs. It is able to display three-dimensional (or more exact: four-dimensional) electron or difference electron distributions (DED's) as hyperareas in space. But not only that: Through "highlighting" by mouse click of special DED's, it is possible to calculate an efg tensor in size and direction with respect to the crystallographic axes and to plot the tensor around the chosen reference point within the unit cell. We will refer to this important feature later.

The program DEDLOT was written within the software frame IDLDE (IDL-Development Environment). It is a very comfortable GUI and includes built-in editing and debugging tools. Having absolved the installing and licensing procedure, IDLDE is started, e.g. in the UNIX X-windows surface of a Mac, just by entering the command "IDLDE" after the prompt. Subsequently, the following window (Fig. 4.4) with the IDLDE-GUI should open (or a similar one; meanwhile, a lot of development was done by RSI Inc.).

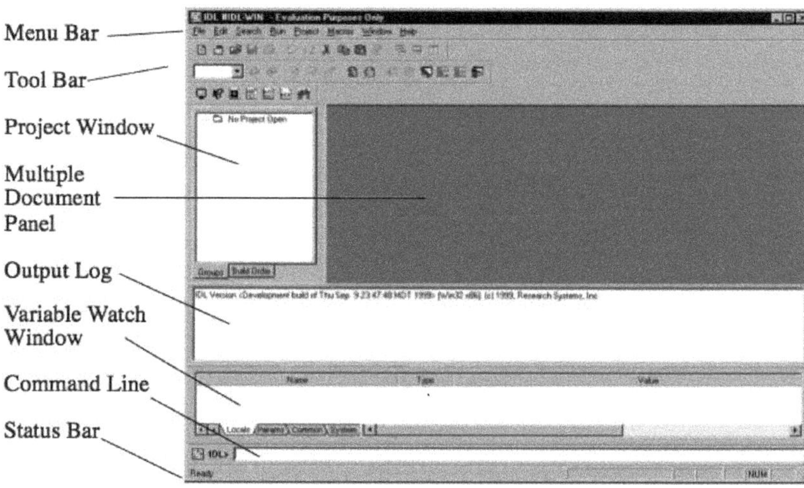

Figure 4.4 Graphical User Interface (GUI) of the Interactive Data Language (IDL) Development Environment (IDLDE) frame by Research Systems Inc™.

The Menu Bar is the top bar of this GUI and contains several options. Selecting one of them creates a brief description in the Status Bar, the last line of the GUI. The Tool Bar contains several icons; placing the mouse cursor over one of them will cause again a Status Bar description. Clicking on one of them (=IDL command) displays the command in the Output Log subwindow. The Project Window on the left hand side of the GUI lists the current projects, e.g. dedlot.pro (.pro is the extension for "projects"). The Multiple Document Panel, e.g., shows the IDL Editor, by which applications are created—the contents of, e.g., dedlot.pro can be seen here. The Command Line is a prompt where single commands may be put in and immediately executed. The Output Log, as the notation is inferring, gives output messages from IDL, e.g., that an error has occurred, a.s.o. In the Variable Watch Window, the timely development of variables can be followed by the user—this is convenient from reasons of easy debugging. The IDL Development Environment as well as IDL itself may be left with "exit" given in the Command Line.

In order to understand the rough mode of operation of our program DEDLOT, which is the most important part of the DEN, it is best to see the GUI of DEDLOT working by help of several screenshots.

By selecting "Project" in the Menu Bar and choosing one from a list (e.g. DEDLOT.pro) and subsequently selecting the "Compile" option, the project file is translated into machine code. This has to be done two times, after that, by selecting "Run", the program will start.

A new window appears with a Menu Bar containing the options "File" and "Options" and the Switches "Settings", "Tools" and "Slots" (Fig. 4.5).

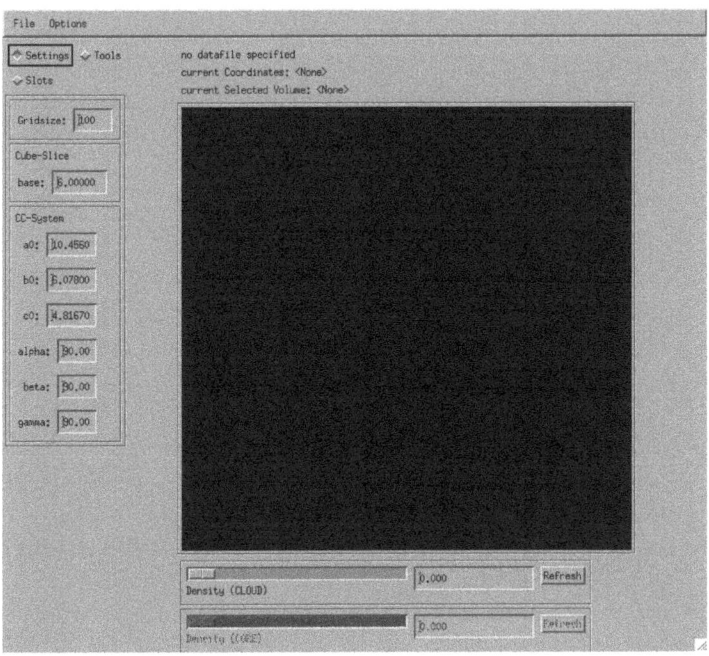

Figure 4.5 GUI of the DEDLOT program for the 3D display of difference electron densities (DEDs).

On the left hand side of the DEDLOT Window, we have several small windows with the possibility of direct input (by clicking into the relevant window):

(1) Gridsize: This is the resolution of the desired cube as given in EVOX; gridsizes of 100, 150 or 200 are usual. With more the abilities of the program may be exceeded.
(2) Cube Slice: This window denotes the edge length of the desired cube (chosen in EVOX) in units of Ångström.
(3) CC-System: The relevant lattice parameters of the considered crystal sample may be put in here according to the values given in EVOX—standard values are provided by the program for synthetic fayalite as dummies.

Below the display window (black), we find a slider for "Density (Cloud)" and again a small window, where the density value is displayed. It may also be directly addressed with a rational number. The button "Refresh" allows to regenerate or actualize the Display Window after a change. There is a second unit of this kind at the bottom of the frame: It contains the same possibilities as above for a "core file" displaying the positions of the different ions in the Cube. This file can also be created by EVOX, taking the calculated structure factors F_{cal} as a basis for the Fourier inversion. Above the Display Window, we find three lines with information about the chosen datafile, a.s.o.

The switch "Tools" (top left) changes the options at the left hand side of the Frame (Fig. 4.6).

There are two general options: "Multiselect" and "Clipping":

(1) Multiselect: By a small window with a slider the diameter (not radius) of a sphere can be defined (in units of Ångström), in which the relevant DED's are highlighted (e.g., for the calculation of an efg from DED's within a certain diameter around the reference point/centre ion of the Cube). To achieve this, at least one DED must manually be highlighted beforehand. A wire-frame sphere in dark blue appears—by the switch "Grab" all the DED's within the sphere can be highlighted—this can last a while since it is no trivial task for a program to find virtual hyperareas and to decide whether they are inside or outside a 3D-boundary.
(2) Clipping: The same as above, only for different emphasis in the direction of the Cartesian coordinates X, Y, Z. The slider gives the value for each direction separately. In fact, we did not use this possibility.

The switch "Slots" (top left) again changes the options at the left hand side of the Frame (Fig. 4.7).

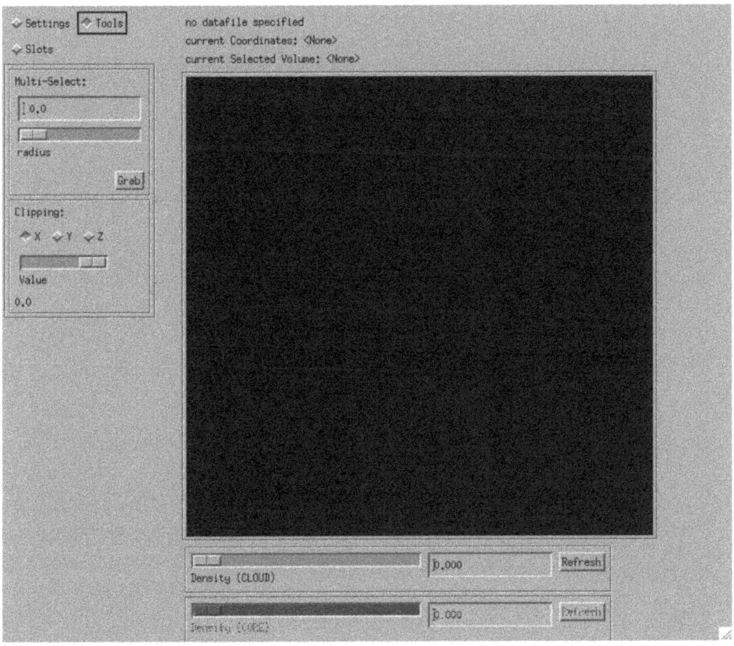

Figure 4.6 GUI of the DEDLOT program for the 3D display of difference electron densities (DEDs), switch "Tools" (top left).

There are again two general options: "Slot Selection" and "Color-Opacity-Settings".

(1) Slot Selection: The relevant button will change the colour slot upon clicking. For each slot we may define another basic colour and another opacity—the latter is the opposite of "transmissivity".

(2) Color-Opacity-Settings:

   (a) By the slider "Opacity" we can determine, how "intense" or better "opaque" a chosen colour should be. In fact, this is the fourth dimension of our later DED images. By choosing a special colour and opacity of a certain DED we may influence the impression of the image in a very sophisticated manner: The inner regions of a DED can be displayed by a "strong" colour and high opacity; by smart application of the Slider "Density (Cloud)" we may show the simultaneous outer regions in a less intense colour and a lower grade of opacity: By these measures a nice grading of the different regions of a DED cloud can be achieved in order to enhance the "3D feeling" of the viewer.

(b) By the button "Color-System" and subsequent pop-up window we can choose, whether we might prefer the common RGB system (red, green, blue) or some other colour system (e.g. CMYK—cyan, magenta, yellow, ...).

(c) The sliders determine the fraction of each special colour within the total displayed colour (e.g., red 100%, green 100% and blue 0% will give yellow). It should be emphasized that the chosen colour of each "Slot" dominates the preceding one—the colours are accumulated. Hence this tool should be applied carefully to obtain the desired effects.

(d) EFG-Size: A bit out of place, this slider controls the relative size of the efg tensor ellipsoid with respect to the surrounding DED's. Sometimes the calculated value of the efg main component is too low to be displayed—the slider compensates this shortcoming.

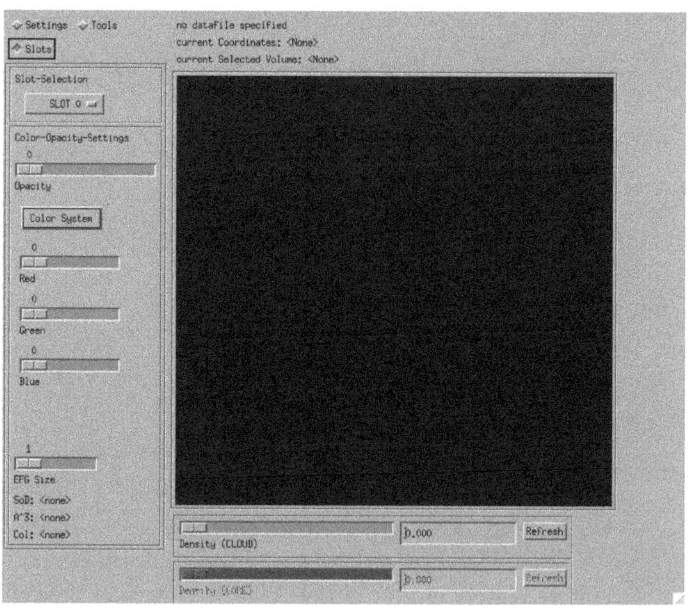

Figure 4.7 GUI of the DEDLOT program for the 3D display of difference electron densities (DEDs), switch "Slots" (top left).

Now let us return to the main options of the Menu Bar. Figure 4.8 shows the pop-up window when choosing the "File" setting:

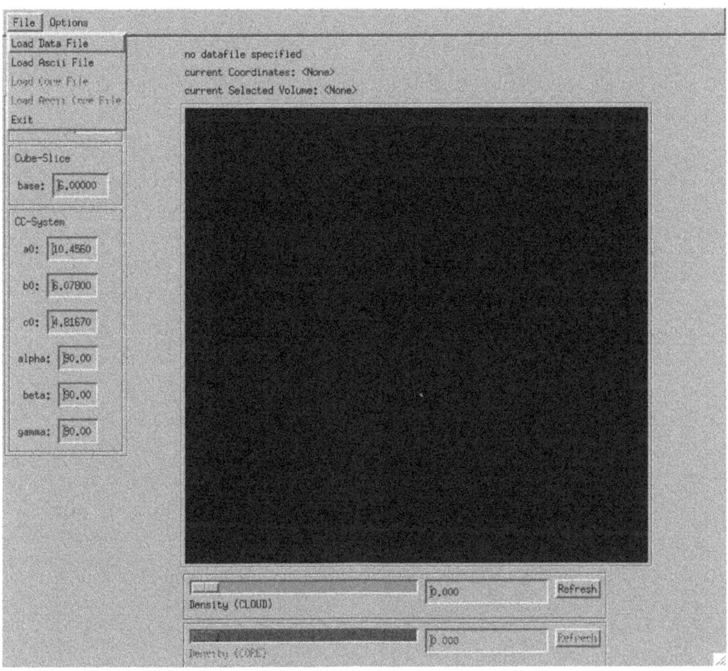

Figure 4.8 GUI of the DEDLOT program for the 3D display of difference electron densities (DEDs), switch "File" (top left) for the choice of the input data file, then "Options".

For example, by clicking on the button "Load Data File" a Browser window opens and let choose the binary File created by EVOX.

By selecting "Options" another pop-up window shows new possibilities to choose: "Core Data On/Off", "Multiselect", "Compute EFG", "Draw EFG" and "Draw EFG from Center of Mass" (Fig. 4.9).

(1) Core Data On/Off: This switch enables us to select a new file containing the calculated clouds of the different nuclei or ion positions in the Cube and to display them properly by the herewith activated slider "Density (Core)".
(2) Multiselect: This switch opens a new pop-up window with the possibility to select a sphere for the collection of the DED's to highlight or something else.

(3) Compute EFG: Here, an efg tensor is calculated from the highlighted DED's:

(4) DRAW EFG: The calculated efg tensor is drawn and plotted as a light blue wire-frame model together with a listing of all important efg data concerning eigenvalues and eigenvectors with the direction angles with respect to the crystallographic axes.

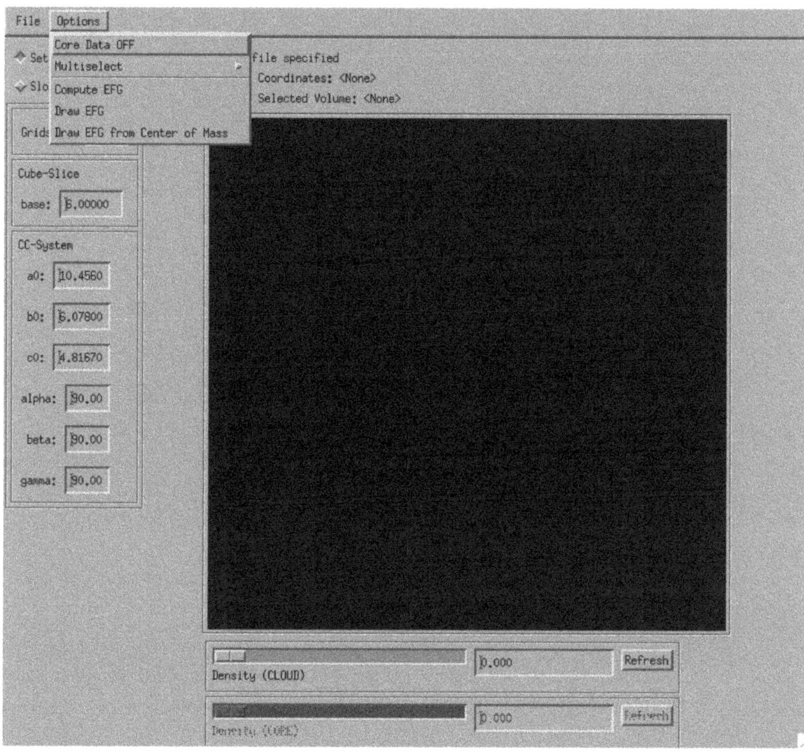

Figure 4.9 GUI of the DEDLOT program for the 3D display of difference electron densities (DEDs), with the five possibilities for choice under the main point "Options" (top left).

In the following sequence an exemplary course of proceeding is displayed with screenshots of every stage.

(1) Selection of the binary data file (in this case DED data for synthetic fayalite based on synchrotron measurements

of Armin Kirfel—on M1 with grid 100 and Cube size 6 Å (Fig. 4.10):

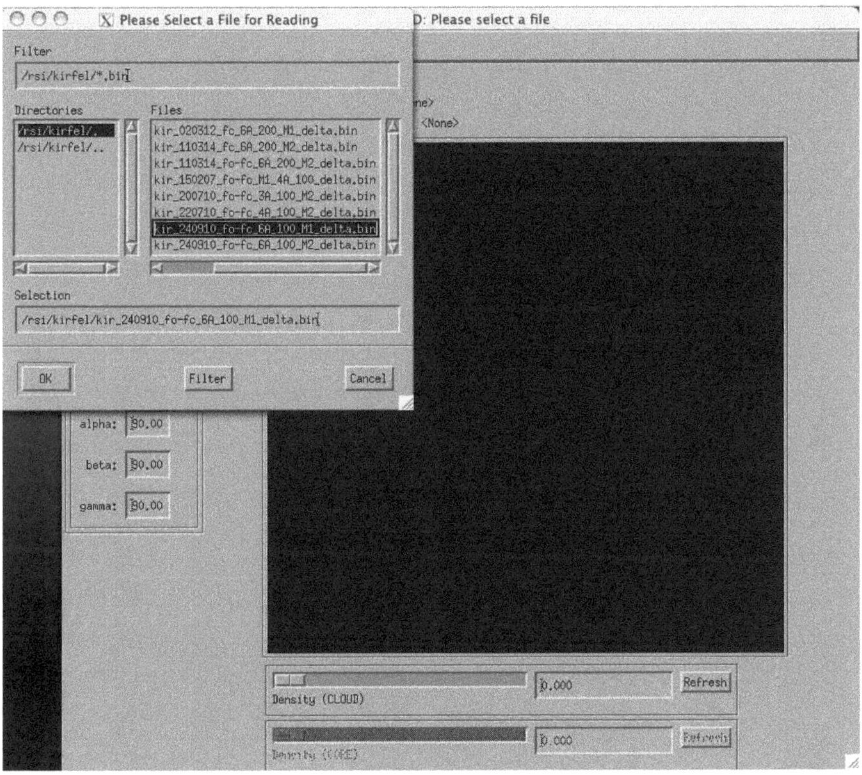

Figure 4.10 GUI of the DEDLOT program for the 3D display of difference electron densities (DEDs), switch "Load Data File" with possibility to browse the available input data files.

(2) The image resulting from these DED's with a Density of 0.179. This value is just selected by experience (Fig. 4.11). It might also be called "render size": It is only reasonable to show a DED image up to a certain render size; below the value too few details are visible, above too much. The corners of the Cube are denoted by small tripods. The tripod at the left corner is coloured in red, green, blue in order to denote the hierarchy of the associate lattice constants *a*, *b*, *c* (SG Pnma), respectively (red = biggest value...).

(3) Some of the DED's are highlighted by clicking on them with the mouse cursor. This is just done for illustration without any physical significance (Fig. 4.12).

(4) Figure 4.13 is induced by the previous clicking on "Compute EFG". Integrating over the previously highlighted DED hyperareas and by help of a point charge algorithm an efg matrix is set up and diagonalized by the Jacobi method. The following listing comprises the Jacobi matrix, eigenmatrix and eigenvalues and eigenvectors as well.

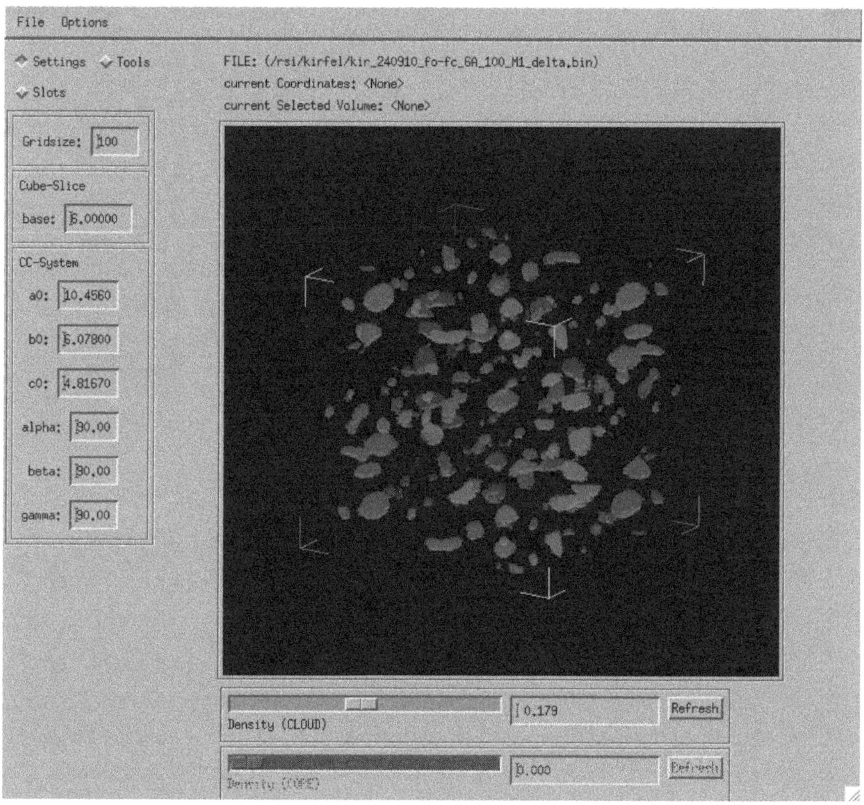

Figure 4.11 GUI of the DEDLOT program for the 3D display of difference electron densities of synthetic fayalite; DEDs within a 6 Å cube around the M1 position 0.5, 0.5, 0.5, reduced by a render size of 0.179 (adjustable slider below the DED window).

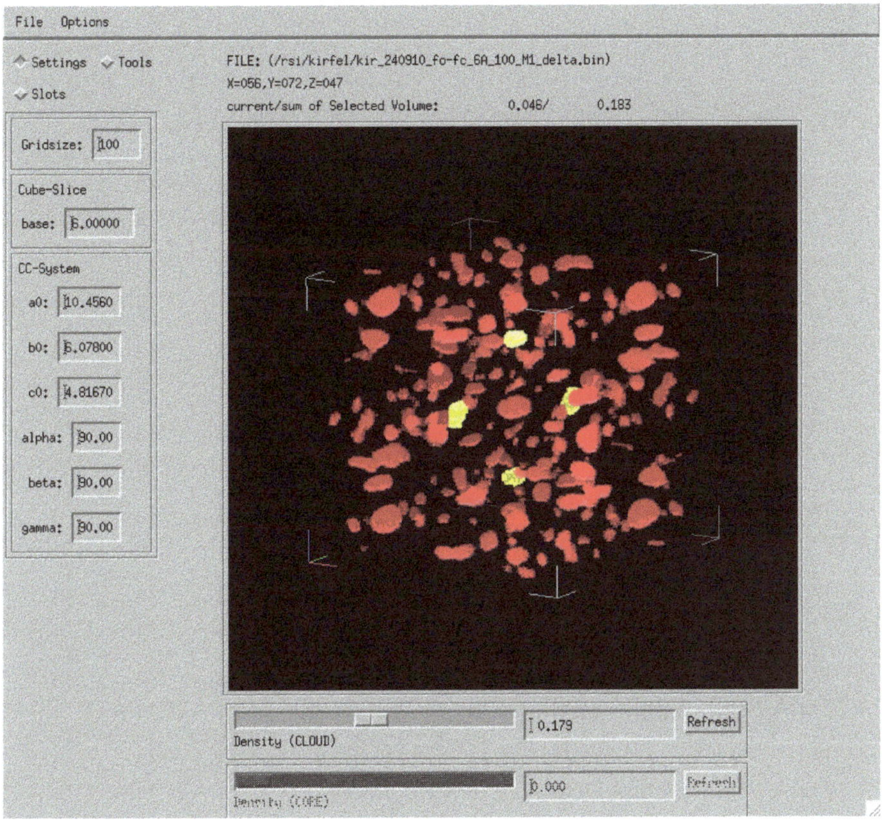

Figure 4.12 GUI of the DEDLOT program for the 3D display of difference electron densities of synthetic fayalite; DEDs within a 6 Å cube around the M1 position 0.5, 0.5, 0.5, reduced by a render size of 0.179, several DEDs selected and highlighted in yellow.

(5) Under "Options" the command "Draw EFG" is selected and under "Tools" the efg size enhanced to "100" by the slider. Figure 4.14 shows the DED window of this procedure: The most important quantitative part is always in the middle of the listing displayed, as shown in Fig. 4.13. Again, there are different eigenvalues and eigenvectors for the efg of the highlighted DED's and the angles [°] of the 3 efg axes with respect to the three crystallographic axes (= 9 values). The Program even calculates the hierarchy of the

efg vectors with the convention $|V_{zz}| \geq |V_{yy}| \geq |V_{xx}|$ (main component of the efg being largest). At this point, it should be confessed that sometimes this hierarchy is not obeyed correctly by the Program—a careful inspection by manual recalculation of the components with respect to their absolute values has soon cleared up a possible error.

```
IDL>
mittelpunkt:        50.0000       50.0000       50.0000

  -0.00039287368       0.0000000         0.0000000
   0.00018711921    0.0012780421         0.0000000
   8.1097820e-05  -0.00041729389   -0.00088516850
Jacobi Matrix:
  -0.00039287368   0.00018711921    8.1097820e-05
   0.00018711921    0.0012780421   -0.00041729389
   8.1097820e-05  -0.00041729389   -0.00088516850
JacobiSpur:    -1.0550566e-10
Diagonalisiert:=Eigenwerte in Zeile     -0.00038673579  -0.00098526846     0.0013720041
DiagSpur:    -1.0550565e-10
Eigenmatrix ohne multipliziert mit Eigenwert * Umrechnungsfaktor
     -0.97657028       0.057806154       -0.20728951
     -0.19275803        0.19330648        0.96201712
    -0.095680917       -0.97943405        0.17763476
------------------------------
Eigenwerte:     -0.00038673579  -0.00098526846     0.0013720041
Eigenvektoren (zeilenweise)multipliziert mit Eigenwerte:
   0.00057253861  -3.3890296e-05   0.00012152863
   0.00028790815  -0.00028872734   -0.0014368926
  -0.00019900669   -0.0020371243   0.00036946245
max_norm:      0.0020798994
längster Vector an Position:         2
Angle (EFGxx...A)          :         12.427186
Angle (EFGxx...B)          :         93.313891
Angle (EFGxx...C)          :         78.036438
Angle (EFGyy...A)         :        78.886212
Angle (EFGyy...B)         :        101.14580
Angle (EFGyy...C)         :        164.15778
Angle (EFGmax...A) [betaA]:          95.490507
Angle (EFGmax...B) [betaB]:          168.35982
Angle (EFGmax...C) [betaC]:          79.767974
------------------------------
mapping:          0.99998096  -0.00019491392    3.5350506e-05
  -0.00019491392       0.99800477   0.00036186410
    3.5350506e-05   0.00036186410      0.99993437
Angle (Vxx...PIx) [alpha_pa]:          12.422239
Angle (Vxx...PIy) [alpha_pb]:          101.16837
Angle (Vxx...PIz) [alpha_pc]:          79.788807
% Compiled module: ORB__DEFINE.
ellirad:                  1

IDL> ▌
```

Figure 4.13 Listing of the calculated efg results (eigenvectors and eigenvalues, Eulerian angles) based on the highlighted DEDs of Fig. 4.12.

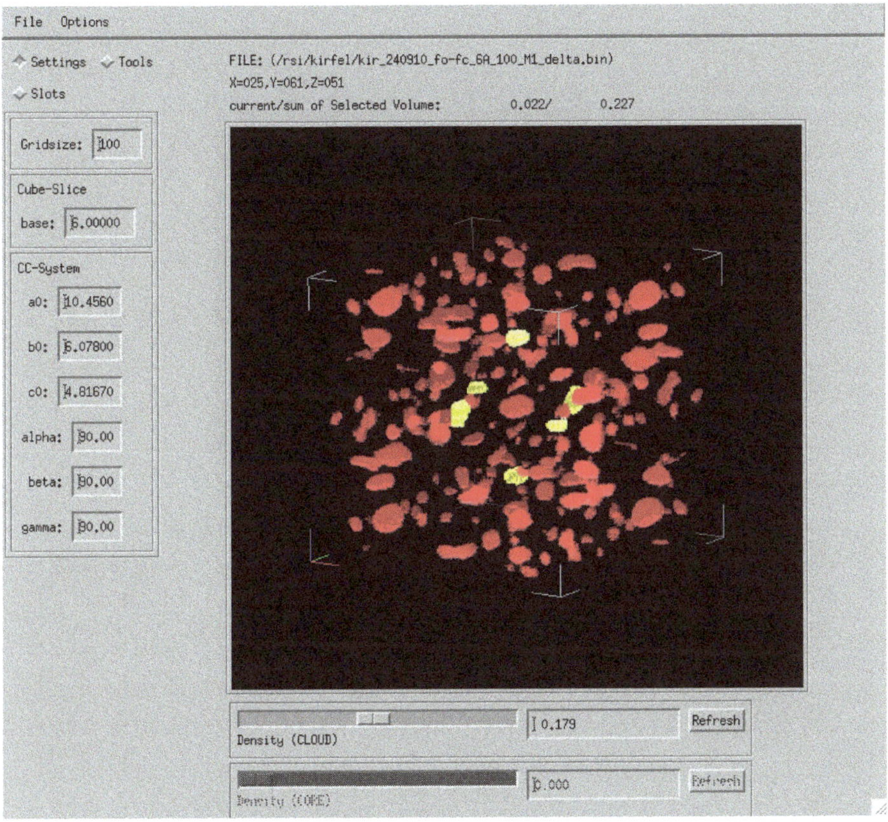

Figure 4.14 GUI of the DEDLOT program for the 3D display of difference electron densities of synthetic fayalite; DEDs within a 6 Å cube around the M1 position 0.5, 0.5, 0.5, reduced by a render size of 0.179, a probable octahedral distribution around the central iron ion is selected and highlighted in yellow.

(6) Simultaneously with the listing an image of the calculated efg as blue wire-frame model within the highlighted DED's (as well as the other ones) is provided (Fig. 4.15).

(7) By the option "Tools/Multiselect" a slider is activated and, in this case, put to 6.4 Å diameter. Immediately, a dark blue boundary sphere (also as a wire-frame model) is displayed (Fig. 4.16).

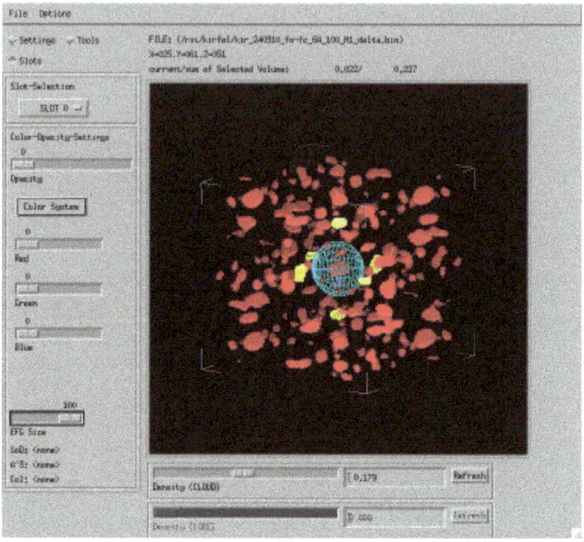

Figure 4.15 GUI of the DEDLOT program for the 3D display of difference electron densities of synthetic fayalite; DEDs within a 6 Å cube around the M1 position 0.5, 0.5, 0.5, reduced by a render size of 0.179, several DEDs selected and highlighted in yellow. The efg due to these DEDs is calculated and plotted as a light blue wire frame model around the M1 iron ion.

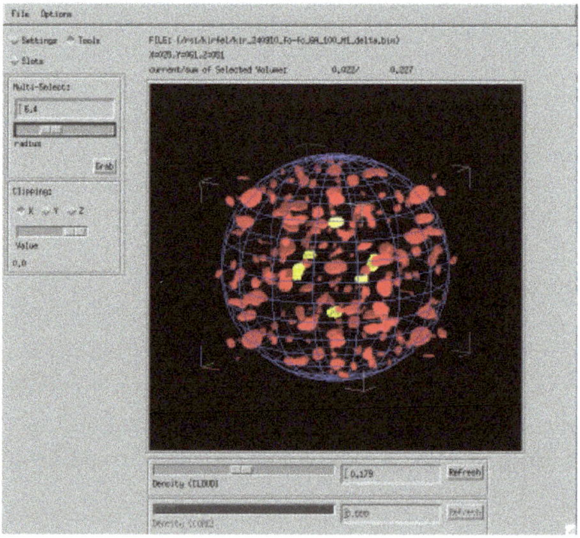

Figure 4.16 Option to define a dark blue wire frame sphere with a certain diameter (slider!) around the chosen central ion in order to collectively select the DEDs without the necessity to perform it one-by-one.

(8) By the button "Grab" all the DED's within the dark blue sphere are highlighted (Fig. 4.17). By this, it is possible to take a whole multitude of DED's for the calculation of the efg. We did this in the case of synthetic fayalite with the reference point on M1 (at 0.5, 0.5, 0.5). The relevant result is shown in Section 5.6.

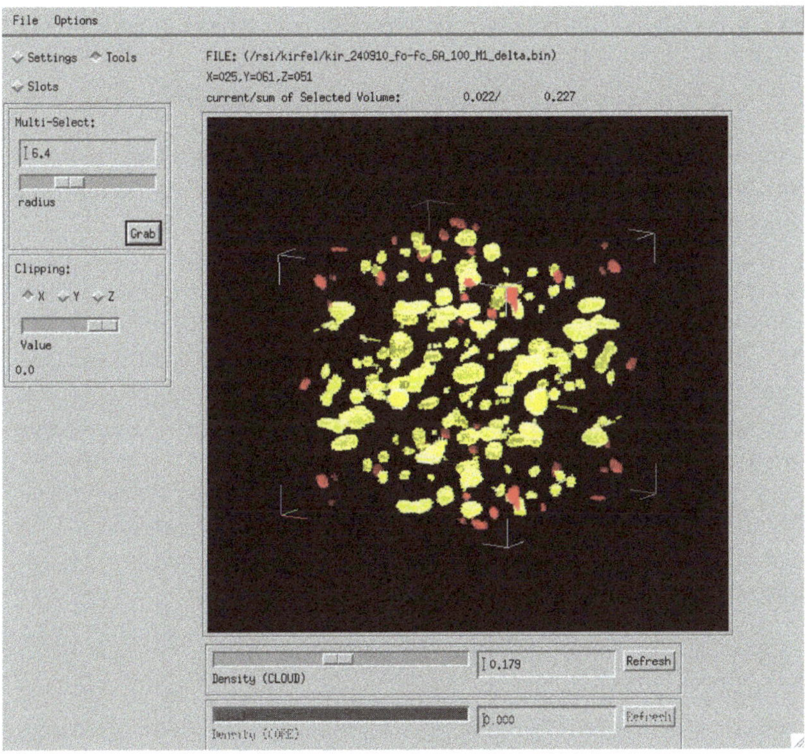

Figure 4.17 Pushing the button "Grab" selects all the DEDs within the dark blue wire frame sphere defined in Fig. 4.16 automatically. By this tool it is not likely to "forget" selecting any of the desired DEDs.

4.3.5 The Routine to Identify Series Termination Errors ("Fourier Ghost Buster System")

From the theory of Fourier series, it is well known that insufficient numbers of Fourier components will lead to series-termination errors preferably at locations where singularities might occur. The general condition to establish a Fourier series from an arbitrary function is that this function need not have no singularities at all, but that it is “partly smooth”. This means that the considered function should allow to be approximated over large distances by periodic functions like sines or cosines with different amplitudes and phases. Considering again that famous example in the 1D range, a rectangle, we at once might understand what is meant: The horizontal sections of the rectangle are well approximated by these trigonometric functions, the sum wave “wiggles” better and better (with decreasing amplitudes of the oscillations) around the horizontal lines and arrives at a comparably well reproduced, rather straight line with increasing number of Fourier components. At the boundaries of the “partly smooth” lines, at the location of the vertical singularities, however, the wiggles intensify but at the same time become narrower.

In two dimensions, however, the deficit of a sufficient number of Fourier components lets arise strange contours in the surroundings of the “real” peak maxima, where the former are not based on existing structures of the original function, but are merely artefacts. Therefore they are in this case not only called “series-termination errors” (having the same cause as the above-mentioned ones, though) but, more ostensibly, “Fourier ghosts”.

In three or, better, four dimensions—our case here—these “Fourier ghosts” are hyperareas which only pretend to be existing difference electron densities, but do not exist in physical reality.

The question now is: How is it possible to identify those artefacts? This is eminently important, since the derivation of a highly reliable efg from DED’s is dependent on the correct selection of the latter. Including non-existing DED’s would change perhaps the absolute value of the efg not so severely, but the direction to a large extent. It will be almost impossible then to avoid series termination errors at all (due to the limited amount of measured structure factors), but it is possible to avoid the artefacts by ruling them out for the calculation of the efg.

This is done by our special selection program (therefore named “SELEC”), otherwise—and a bit ironically referring to the homonymous US movie—(Fourier-) ghost buster routine.

It is based on the observation mentioned above that the less Fourier components we have, the more and intense “ghosts” we have to deal with. The program “SELEC” now reviews the available structure factors up to a certain limit that is expressed by sineθ/λ and selects all these ones, which are within the limit. The low values of sineθ/λ will produce structure factors more close to the nucleus; the higher values, however, will also regard the "outskirts" of the atom. By this, we will select differently voluminous data sets. The latter are successively put into DEDLOT and the effects observed with the following results:

(1) The local symmetry of the crystallographic site is decisive for the formation of Fourier ghosts: The lower the symmetry, the less likely we will suffer from artefacts.

(2) In the case of the M1 position of synthetic fayalite with low point symmetry −1 the different data sets did not produce considerable artefacts—we therefore could take all DED's within a sphere of 6 Å diameter around the reference nucleus (M1 at 0.5, 0.5, 0.5).

(3) In the case of the M2 site with comparably high point symmetry *m*, we observed artefacts—we followed them on the DEDLOT images for the different data sets and could therefore rule them out for the calculation of the relevant efg. Consequently, the so-derived efg provided good results, whereas the "false" one with the artefacts included produced an erroneous efg. Comparison was done by help of the measured efg values and those calculated by DFT.

Hence, we conclude that the program SELEC is highly reliable, at least in the case of our sample.

Chapter 5

Application of the DEN on a Representative Example (Synthetic Fayalite Fe_2SiO_4)

For a researcher like me, often at the beginning of the scientific career, certain decisive questions might arise which influence the course of the whole later scientific life:

(1) Shall I deal with only few samples (more or less important) and, at the same time, many methods in order to get a possible consistent image of the underlying physics but take the risk to publish not so many papers?

(2) Shall I deal with many samples and a standard method to become a specialist in the application of the method and shoot out one publication after the other?

(3) Shall I deal with a lot of samples and a lot of methods (being well skilled in all of them) and also publish a lot of papers?

Besides the fact that the decision is highly dependent on outer circumstances (size of the research group, equipment of the lab, fundraising a.s.o.), the third aim is difficult to achieve, if not utopic.

In any case, I might confess here that I have been trying to act according to the first point above. I started with magnetometry

The Difference Electron Nanoscope: Methods and Applications
Werner Lottermoser

ISBN 978-981-4774-01-7 (Hardcover), 978-1-315-19664-0 (eBook)
www.panstanford.com

(susceptibility measurements) and magnetism, then turned towards neutron diffractometry and crystallography, and then Mössbauer spectroscopy/solid-state physics. By all these methods, I did not investigate only one sample, as the precedent tenor of the book might be inferring so far, but "my" main important sample with lots of experimental and theoretical data is synthetic fayalite Fe_2SiO_4 indeed. More pathetically speaking, I would like to claim that I wanted to see the "rainbow in one waterdrop" and not to examine lots of waterdrops to detect the rainbow (this procedure is also possible, though).

Why is fayalite Fe_2SiO_4 so particularly interesting, although it is not technically relevant or a specimen being from other commercial reasons outstanding?

I will try to answer this in the following section.

5.1 Fe_2SiO_4: Description of Its Crystallographic and Magnetic Properties

α-Fe_2SiO_4, or fayalite (its mineral name), is an island silicate belonging to the olivine group of minerals with the general formula $(Mg, Fe)_2SiO_4$ due to its olive-green colour. Fayalite itself, however, is brown due to its contents of Fe^{2+}-ions—the name stems from the volcanic island Faial, part of the Azores (Portugal) where the mineral was detected in 1840. In nature, it only occurs in very small specimen in the mm-range, and bigger ones have been found in the slag of historical smeltings. This was rather surprising, since even nowadays it is rather difficult to produce larger synthetic fayalite specimen in the cm-range. This is due to the fact that a very low oxygen pressure has to be maintained during synthesis in order to prevent the Fe^{2+} ions from getting transformed into Fe^{3+}, which would create the Fe^{3+} analogon of fayalite, the comparably late detected laihunite.

However, mostly Japanese experts, in the course of producing all kinds of crystals, preferably semiconductors for the computer technology, managed to establish and handle the difficult boundary conditions for the fabrication of large fayalite specimen in the cm range by the Czochralski-method. This procedure, by the way, has been detected due to a strange serendipity science often has benefitted from:

While writing, the crystal researcher Czochralski put his pen accidentally into a container of liquid tin instead of the inkpot. By slowly removing the pen, he pulled a sort of tin fathom out of the container—on closer inspection, he perceived that the fathom was a single crystal! This hazard has revolutionized the technique of the production of large crystals since even nowadays the comparably giant Si-crystals of several centimetres in diameter and several tens of centimetres in length for the famous Si-wafers as a basic material of computer technology are drawn according to this procedure. Also the large fayalite crystals for our experiments have been produced with this method in the lab of S. Hosoya and his co-workers at Sendai, Japan.

Early structure investigations with X-rays certainly on silicates and, in particular, olivines were performed in the early days of the X-ray method for sure. However, X-ray, neutron powder, magnetometric and Mössbauer measurements were undertaken in the late 1960s. From X-ray data, it was known that fayalite crystallizes in the orthorhombic space group Pnma with four formula units Fe_2SiO_4 per unit cell. The crystallographic structure is represented by Fig. 5.1, where the main important features are graphically shown (CaRIne, Boudias and Monceau, 1989–1998):

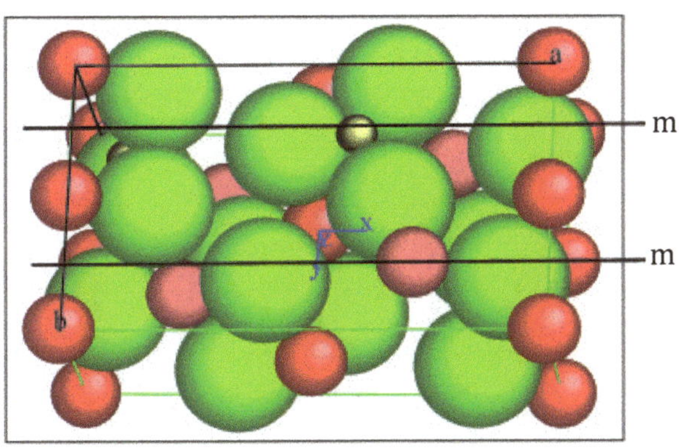

Figure 5.1 Schematic view of synthetic fayalite (SG Pnma) with relevant ions displayed as shaded balls: Fe^{2+} on M1(4a) position with −1 point symmetry: dark red, Fe^{2+} on M2(4c) site with m(irror) symmetry: light red, Si^{4+} on 4c (m): golden, oxygens $O(1)^{2-}$ and $O(2)^{2-}$ on 4c (m): light green and oxygens $O(3)^{2-}$ on 8d (1): light green.

The Fe^{2+} cations are displayed as differently shaded balls, occupying the two non-equivalent crystallographic sites M1 (or 4a according to the General Tables notation) with −1 local symmetry and M2 (or 4c) with point symmetry *m*. The tiny Si^{4+} balls on M2 are the centres of the different oxygen tetrahedra which are crosslinked to two differently distorted octahedra. The oxygens O(1) and O(2) also occupy 4c sites, the O(3) ones the general position 8d. Very accurate X-ray data stem from single crystal measurements by Fujino et al. (1981), the atomic positions and lattice constants by J. W. Bats, published in Lottermoser et al. (2002) and listed in Table 5.1:

Table 5.1 Crystallographic parameters of synthetic fayalite according to the X-ray four circle diffractometer analysis by J. W. Bats

Empirical formula	Fe_2SiO_4		
Formula weight	203.79		
Wavelength	0.71073 Å		
Crystal system, space group	orthorhombic, Pnma		
Volume	306.30(10) Å^3		
Formula units *Z*	4		
Calculated density	4.419 kg/m^3		
Absorption coeff.	9.690/mm		
Extinction coeff.	0.087(4)		
Unit cell dimensions	α = 10.4597(17)Å α = 90°	b = 6.0818(15)Å β = 90°	c = 4.8150(7)Å γ = 90°
Reflections coll./unique	20712 1575		
Goodness-of-fit F^2	1.156		
Final R's ($I > 2\sigma(I)$)	$R1$ = 0.0223	wR2 = 0.0708	
R's total	$R1$ = 0.0249	wR2 = 0.0719	
Atomic coordinates	*x*	*y*	*z*
Fe(1)	0.00000	0.00000	0.00000
Fe(2)	0.28003(2)	0.25000	0.98599(4)
Si	0.09745(3)	0.25000	0.43085(8)
O(1)	0.09174(9)	0.25000	0.76800(20)
O(2)	−0.04688(9)	0.25000	0.29110(20)
O(3)	0.16580(6)	0.03601(11)	0.28842(14)

From Fig. 5.1 it may be seen that the Fe^{2+} cations on M1 form long rows of ions; the Fe^{2+} cations on M2, however, constitute zig-zag lines within higher symmetrical oxygen octahedra with internal mirror planes. Each M2 cation opposes two M1 ions thus forming peculiar triangular plates—this feature is supposed to cause the difficult magnetic properties of synthetic fayalite at low temperatures:

The early susceptibility and neutron diffraction measurements on powder samples by Nomura et al. (1964) and Santoro et al. (1966) yielded a collinear antiferromagnetic (AF) structure below $T_N \approx 65$ K parallel to **b** (SG Pnma) on M2 and a temperature-dependent canting on M1. Mössbauer measurements on powder samples by Kündig et al. (1967) provided well-refined spectra with two subspectra being attributed to M1 and M2. In those days, however, it could not be explained why the angle parameters θ between the internal magnetic field $H(0)$ and the main axis of the efg on M1 should remain independent of temperature. Also, a supposed second transition at $T \approx 23$ K remained unexplained. The neutron diffraction measurements by Müller et al. (1982), however, could solve the alignment of the magnetic moments dependent on temperature for M1 and M2—the results are shown in Figs. 5.2 and 5.3.

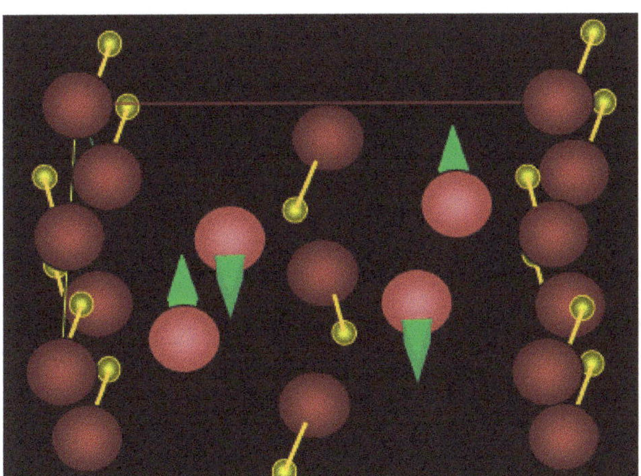

Figure 5.2 Nearly collinear antiferromagnetic spin structure of the Fe^{2+} ions on M1 (dark red balls, yellow pins) and collinear AF structure along **b** (SG Pnma) of the Fe^{2+} ions on M2 (light red balls, green arrows) of synthetic fayalite at $T = 50$ K according to neutron diffraction data.

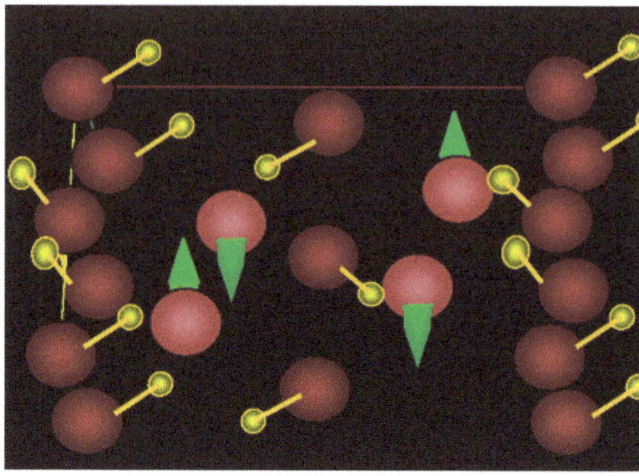

Figure 5.3 Canted antiferromagnetic spin structure of the Fe^{2+} ions on M1 (dark red balls, yellow pins) and collinear AF structure along **b** of the Fe^{2+} ions on M2 (light red balls, green arrows) of synthetic fayalite at $T = 10$ K according to neutron diffraction data.

However, the second transition could not be verified, and also the discrepancy of the Mössbauer results remains unexplained.

Parallel to our own experiments on synthetic fayalite, theoretical studies were performed with classical approaches, first by Ballet et al. (1989) and, very successfully, by Ehrenberg and Fuess (1993).

We then managed to solve the "Mössbauer problem" by our measurements on single crystal specimen at many different temperatures below T_N: It turned out that the temperature dependence of the internal magnetic field (and thus the one of the magnetic moments) was hidden in a parameter which was given a constraint in the early work of Kündig et al.; it was fixed at 90°. However, in those early days of hyperfine spectra fitting, the refinement was done by nomograms; thus it was almost impossible for them to detect this shortcoming. Later on, we could also explain the mysterious "second transition" —our Mössbauer results revealed (Lottermoser et al., 1996) that it is a collective sudden change of canting around $T \approx 23$ K, leading to strong effects in the susceptibility $\chi(T)$ and specific heat $c_P(T)$ (Robie et al., 1982) as "collective" methods, but not in neutron diffraction. For "economic" reasons only few data sets at selected temperatures and few reflections dependent on temperature can be measured in the available facility time; so such peculiarities can be easily overseen. This was explained in detail in our publication Lottermoser et al. (1996). Up to then, the following explanation of the magnetic properties could be established:

The particular crystallographic structure of fayalite gives rise to the difficult (dynamic) AF structure below $T_N \approx 65$ K. The Fe^{2+} ions show paramagnetism above T_N, below the Néel temperature the couplings of the M2 cations via superexchange overcompensate the temperature motion—the crystal receives a basic antiferromagnetic spin structure parallel to **b**. With increasing strength of the coupling with decreasing temperature, the M1 spins react more and more to the increasing internal magnetic field provided by the M2 moments—they form a highly frustrated spin system, as it is expressed in the relevant terminology.

With increasing coupling forces to the dominant M2 spins, the canting increases more and more in order to compensate for the local 2:1 arrangement of the M1 and M2 cations, respectively.

At this point, we intended to investigate the cause of these properties and the superexchange pathways in much greater detail and started our DFT, Mössbauer spectroscopic and synchrotron diffraction investigations, which led to the development of the DEN.

5.2 Derivation of the Experimental efg by SCMBS

The course of the general determination of an efg according to the point charge model has already been described in the earlier chapters of this book. In the following section, however, it shall be explained how the experimental efg is derived by SCMBS (single crystal Mössbauer spectroscopy) and which preparations are necessary to obtain it.

The decisive step among these is to evaluate whether the crystal sample considered is a single crystal or not. In the case of transparent crystals, polarization microscopy is the method of choice to answer this question. The crystal will be brought on the object table between crossed polarizers and the table be turned. In the field of view, the crystal must then show a "uniform extinction" when the table is turned. This means that preferably the whole crystal (or, in less favourite cases, parts of it) has to brighten or darken in the whole and not subsequently. In the second case, when only parts of the crystal show this behaviour, these parts might be "cut out" by a fathom or diamond saw in order to obtain at least a single crystalline piece of the original specimen.

For opaque crystals, an X-ray analysis should be performed: Multiple crystal conglomerates or otherwise disturbed samples commonly show multiple reflections lying very close to each other in a Laue diffractogram. For example, twinning can easily be detected by double reflections—it is not advisable to take such specimen for the further measurements since the later efg will be severely doubtful, because the obtained angle directions will be ambiguous.

Provided we have got a nice single crystal of, say, $6 \times 6 \times 6$ mm^3 —such sizes are very often necessary, since in case that the crystallographic axes are severely tilted inside of the bulk, it is desirable to have space for the cut-off. For SCMBS, sizes of $4 \times 4 \times 0.1$ mm^3 are minimum, because the incoming Mössbauer γ-rays need a certain area in order to be reasonably absorbed.

The same is valid for neutron diffraction, whereas for X-ray and, in particular, synchrotron radiation, small spheres with diameters of 0.2 mm and less are preferable. The latter can be prepared in a classical ball mill—the general sketch of the device is shown in Fig. 5.4.

The inner walls of the device are covered with abrasive paper of different grain sizes. By leading a stream of pressurized air through the device, the inserted crystal pieces (preferably previously brought into an isometric shape) will be blown alongside of the abrasive walls and then be etched to tiny spheres (provided the crystalline anisotropy be not to severe).

This procedure can last up to several days (depending on the hardness of the crystal); small grain sizes of paper are, in particular, used in the end of the treatment, where the spheres can be hardly detected by eye view. The qualitatively best ones of these samples are then glued to glass fibers and put into small brass cylinders which are mounted on goniometer heads (cf. Section 3.3.2.1, Fig. 3.32).

The crystals can then be measured on an X-ray or synchrotron diffractometer according to the prescriptions given above. The samples for neutron diffraction or SCMBS are the plates mentioned above (with thicknesses of 0.5–1 mm and about 0.1 mm, respectively).

Figure 5.4 Ball mill for the production of tiny crystal spheres for X-ray diffraction purposes. Parts drawn in dark yellow are made from brass, the raw material and several crystal spheres after grinding with the help of a pressurized air stream as green items.

How these samples are prepared?

For our former measurements of fayalite, a larger original specimen was examined by the BPC (cf. Section 3.3.2.1). In this section, it is demonstrated how the directions of the different lattice constants have been determined. Afterwards, the goniometer head with the oriented crystal was mounted on a special device (Fig. 5.5) and glued to a glass plate with the help of an epoxy glue or heated wax:

The crystal was subsequently cut with a fathom saw (the fathom covered by diamond dust) using a special pivotable support for the glass plate and the crystal, where a counter weight could be adjusted in such a way that the crystal was successively cut by its own weight. This device was used at the Institute of Crystallography at Frankfurt, Germany.

For our purposes now we are using a commercial low-power low-cost X-ray device with molybdenum tube (35 kV, 1 mA) from

PHYWE Ltd., Göttingen, Germany (XR-4.0 X-ray expert unit), in connection with a goniometer head with big arcs (HUBER Ltd., Rimsting, Germany) and later installed motor drives for the sample tilting (Fig. 5.6):

Figure 5.5 Device for the fixing of crystals (preparation of cut) after orientation perpendicularly to selected lattice vectors on an X-ray camera. The goniometer head with the crystal can be seen on the metal adapter below the glass plate.

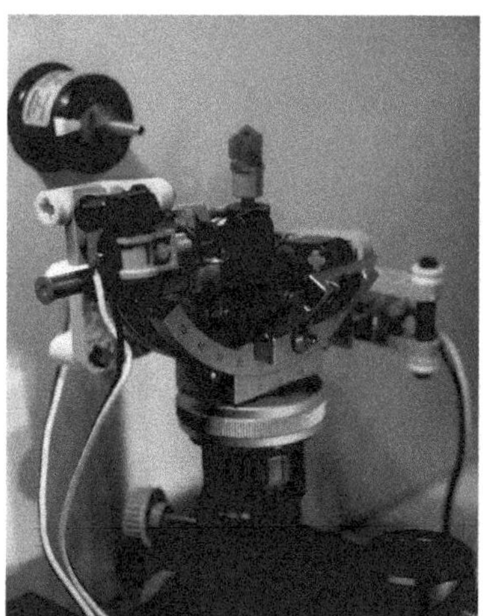

Figure 5.6 Photograph of our motorized goniometer head on our X-ray crystal orientation device. The collimator in front of the crystal sample can be seen in the top left background.

The great advantage of these devices, among others, is the freely configurable compartment for the additional equipment. Behind the fluorescent screen, a standard digital mirror reflection camera (CANON EOS 350D) with manual "bulb" and macro settings and 8 Mpixel high-sensitivity CMOS chip has been installed. The camera can be remotely operated and is connected to an external computer with image processing software.

The following sequence of Laue diffractograms (exposure time: around 1800 s) with orientations close to and almost perfectly parallel to the [010] direction of synthetic fayalite, respectively, has been recorded (cf. Section 3.1.2.1) (Fig. 5.7):

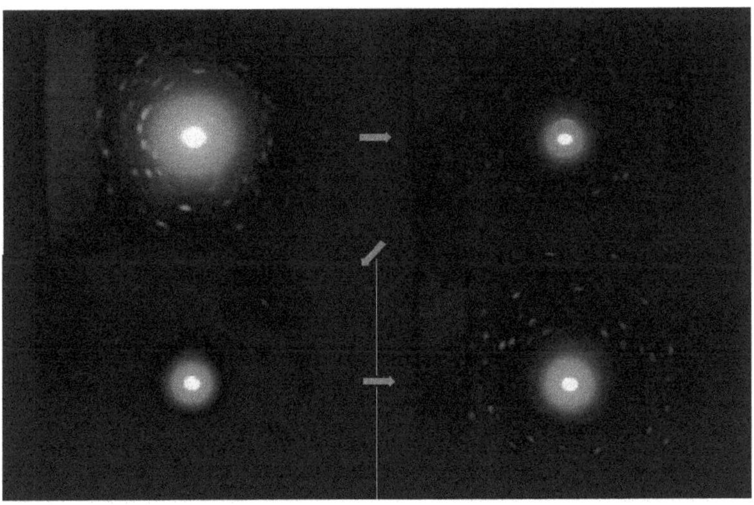

Figure 5.7 Laue pattern of a synthetic fayalite single crystal with the small edge of its plate (b-axis) parallel to the primary beam. Successive better alignment of the crystal axis by adjusting the goniometer head arcs properly. The Laue symmetry mm can be well observed in the last image.

For the evaluation of a complete efg with respect to all three crystallographic axes, however, we need crystal sections in at least three directions perpendicular to each other, since the axes of the efg can be oriented only with two significant angles (β and α) at maximum with respect to the crystallographic axes. So we need another two sections in order to unambiguously align the three axes of the efg with respect to the crystal axes (= 9 angles!).

To reach this aim, we thus have to prepare three crystal sections, in the case of, e.g., fayalite with **a**, **b**, **c** parallel to the **k**-vector of the incident γ-rays.

These sections are prepared as already mentioned above: The original crystal specimen is oriented by X-ray diffractometry and glued to a glass plate. This is mounted on a special turntable with a small diamond saw and an increment of only 0.05 mm at minimum (Fig. 5.8).

Figure 5.8 Milling machine table with *x*–*y* spindle axes. In the centre of the photograph, the bench vise for the glass plate with the crystal can be seen, in the background the circular diamond saw on its motor drive.

By this, a very gentle cut with low risk of destroying the crystal can be performed. A second cut parallel to the first one in order to create a thin plate with a thickness of say, 1 mm, is carefully being carried out. Such a specimen can then be used for neutron diffraction purposes.

Another sample for SCMBS is then separated from the glass plate, put into another one with the help of heated wax and successively polished down to the desired final thickness. For SCMBS, the thickness must be really low (around 0.1 mm), otherwise the crystal "does not open" for the weak X-radiation of the Mössbauer source. Therefore, by mutual measuring and polishing, the minimum thickness will be evaluated where the crystal becomes transparent for Mössbauer resonance. If there is, however, a large background of the spectra visible, this might be a hint for too much incoherent scattering being involved.

This problem can particularly be solved by inserting a thin mask of lead in front of the crystal's thin section so that only the area of the crystal is gated out. The mask with the thin crystal can now be mounted in a conventional absorber ring and measured with SCMBS, preferably for all three cuts mentioned above and different temperatures (cf. Section 3.1.2.1).

The result after the spectra refinement will be a table of the different Mössbauer parameters for each thin section and dependent on temperature. Particularly important are the angle values β and α providing the orientation of the **k**-vector of the incident γ-rays (and thus of **a, b, c**) with respect to the efg axes. These are cited in Table 5.2 as an excerpt of the parameter table from Lottermoser et al. (2002):

Table 5.2 Orientational angles β and α for 3 crystal sections perpendicularly to the three crystallographic axes of synthetic fayalite (M1 and M2) at different temperatures in the moderate temperature region; subscripts denote standard deviations

		k//a		***k//b***		***k//c***	
***T* [°C]**	**Site**	β	α	β	α	β	α
RT	M1	122.9_9	329_3	54.2_9	233_3	21_{12}	1.9
	M2	89.9	−181	-27.2_7	78_{11}	90	90
100	M1	125_1	335_4	55.5_8	240_3	28_4	-392_{31}
	M2	79_2	-206_2	-26.7_7	90	74_3	90
150	M1	122_1	325_4	56.4_8	243_4	24_{13}	1.9
	M2	90.0_x	−181	-25.8_7	90	84_{28}	90
200	M1	120_2	322_5	58.0_8	253_6	30_8	1.9
	M2	89.9_x	−181	-23.7_7	90	102_8	90
250	M1	119_2	317_5	57_1	263	34_9	1.9
	M2	89.9_x	−181	-23_1	90	78_9	108_{12}
300	M1	120_2	321_5	59_2	278_{18}	—	—
	M2	86_{15}	-192_5	-23_2	90	—	—

Now it is necessary to know how the angles β and α can be transformed into the axes of the efg with respect to the crystallographic axes. This is not as easy as it might seem initially.

First, we have to take the angle β between the **k**-vector of the incident γ-rays (which is identical with the crystallographic axes **a, b, c**, respectively) and the main component of the efg

V_{zz} as well as the angle α between the projection of **k** (or **a, b,** c, respectively) into the plane (V_{xx}, V_{yy}), taken from V_{xx} (cf. Section 3.1.2). These angles should be transformed into polar coordinates, e.g., according to the book of Haussühl (1977) the angles β and α correspond to ξ and η according to the notation in this book. From the latter, the three polar coordinates x_1, x_2, x_3 have to be calculated which correspond to the crystallographic axes **a, b, c** with respect to the axes of the efg:

$$x_1 = \sin\xi\cos\eta; \quad x_2 = \sin\xi\sin\eta; \quad x_3 = \cos\xi; \quad r = \sin\xi \tag{5.1}$$

The radius of the sphere shall be equal to unity.

If we wish to get a stereographic projection (s.p.), we have to multiply the former coordinates by a factor 1/(1 + cosξ), which takes into account the fact that a stereographic projection upsets the values x_1 and x_2 with approaching the basic equatorial circle:

$$x_1' = \frac{\sin\xi\cos\eta}{1+\cos\xi}; \quad x_2' = \frac{\sin\xi\sin\eta}{1+\cos\xi}; \quad r' = \frac{\sin\xi}{1+\cos\xi} \tag{5.2}$$

A stereogram is a very convenient method to display vectors with their poles in three dimensions—the stereographic projection has two main properties (cf. Section 3.1.2):

(1) The projections of two vectors onto the equatorial plane intersect with the same angle as do the vectors themselves
(2) All circles on the sphere will project as circles on the equatorial plane

Hence, a stereogram is a well-suited representation of angle and vector properties.

However, we should return to the second task to be solved:

We should not calculate the crystallographic axes with respect to the axes of the efg, but vice versa:

It is highly desirable to see the tensor ellipsoid of the efg with its three main axes within the axes of the unit cell. Hence, we should determine the direction cosines of the values x_1, x_2, x_3 (divided by r) with the three equations:

$$\begin{aligned}
\cos\mu &= x_1/r; \ \mu = \arccos x_1/r \\
\cos\nu &= x_2/r; \ \nu = \arccos x_2/r \\
\cos\xi &= x_3/r; \ \xi = \arccos x_3/r
\end{aligned} \tag{5.3}$$

We will then get a table with the angular values of V_{xx}, V_{yy}, V_{zz} with respect to **a, b, c** (cf. Fig. 3.14).

Normally, this would be the goal of our calculations and has been derived, e.g., for Table 5.4 in our publication Lottermoser et al. (2012) to give the experimental angle components of the efg from SCMBS. Also from our DFT calculations and from the DEN method we get analoguous angles which can be easily compared with each other in order to estimate the error of the different methods.

5.3 Calculation of the Full Quantitative efg by the DFT Method

In the theoretical part of this book (Section 3.2), the density functional theory already has been presented. For our calculations on synthetic fayalite, we are using a special form of it, the spin-polarized self-consistent charge (SCC) Xα method in local spin density approximation (LSDA) derived by M. Grodzicki (1980, 1985). The Xα potential (with α = 0.7), however, is only applied for the evaluation of the two- and three-centre integrals (cf. Section 3.2.1.5), whereas the one-centre integrals are derived from all -electron atomic calculations (relativistic). From the latter, a pseudopotential is created, by which the core electrons enter the two- and three-centre integrals. Therefore, this method might be called ab initio, since it does not contain adjustable parameters. The valence basis set contains 2s and 2p orbitals for O (and F, for the reason see below), 3s and 3p for Si (and Mg, dto.) and 3d, 4s, 4p orbitals for Fe.

Using the LSDA, ionization and excitation energies are evaluated by the transition state procedure given by Slater (1974). Mostly, all orbital energies are then shifted, so that energy differences are not mainly affected—so d–d transition energies are calculated as the relevant orbital energy differences. How hyperfine parameters are determined by use of a valence-electron-only molecular orbital method was also described by Grodzicki et al. (1987).

For the evaluation of the efg tensor V_{ij}, the charge density must be divided into the positive point charges Q_μ^{corr} of the ionic cores and the charge contribution stemming from the valence electrons:

$$V_{ij} = \sum_{\mu \neq 0} Q_\mu^{\text{core}} V_{ij}(R_{\mu 0}) - e_0 \sum_{\mu\nu\alpha\beta} P_{\alpha\beta}^{\mu\nu} \int \varphi_\alpha^\mu (r - R_{\mu 0}) V_{ij}(r) \varphi_\beta^\nu (r - R_{\nu 0}) d^3 r \tag{5.4}$$

The considered iron atom (here: Fe^{2+} on M1 or Fe^{2+} on M2, respectively) is located at R_0, $R_{\mu 0} = R_\mu - R_0$, e_0 is the elementary charge, $\phi_\alpha^\mu(r)$ is the α-th atomic orbital at the μ-th atom, $P_{\alpha\beta}^{\mu\nu}$ the bond-order matrix and the efg components $V_{ij}(r)$ given as

$$V_{ij}(r) = \frac{1-\gamma(r)^3}{r^3}\frac{r_i r_j - r^2\delta_{ij}}{r^2} \tag{5.5}$$

$\gamma(r)$ is the Sternheimer shielding function stemming from the polarization of core electrons (of Fe) by charges outside the core.

If we assume a certain basis set of atomic orbitals (cf. Section 3.2.1.4), the efg can be split into three different contributions:

(1) Valence contribution with both atomic orbitals at the Fe-site

$$V_{ij}^{\text{val}} = -e_0 \sum_{\alpha\beta} P_{\alpha\beta}^{\infty} \int \varphi_\alpha^0(r) V_{ij}(r) \varphi_\beta^0(r) d^3r \tag{5.6}$$

The radial part of the remaining one-centre integral is proportional to $<r^{-3}>$ $(1 - R)$ with the Sternheimer anti-shielding constant $R = 0.075$ (Lauer et al., 1979). The expectation value $<r^{-3}>$ is determined with relativistic atomic calculations dependent on the valence shell occupation numbers.

(2) The covalence contribution with one atomic orbital at the Fe site and the other at a ligand atom might also be very important:

$$V_{ij}^{\text{cov}} = -e_0 \sum_{\nu\alpha\beta} P_{\alpha\beta}^{0\nu} \int \varphi_\alpha^0(r) V_{ij}(r) \varphi_\beta^\nu(r - R_{\nu 0}) d^3r \tag{5.7}$$

This term describes the anisotropy of the electronic charge distribution in the Fe-ligand bonds

(3) The ligand contribution separated with both atomic orbitals at ligand atoms:

$$V_{ij}^{\text{lig}} = \sum_{\mu\neq 0} Q_\mu^{\text{core}} V_{ij}(R_{\mu 0}) - e_0 \sum_{\mu\nu\alpha\beta} P_{\alpha\beta}^{\mu\nu} \int \varphi_\alpha^\mu(r - R_{\mu 0}) V_{ij}(r) \varphi_\beta^\nu(r - R_{\nu 0}) d^3r \tag{5.8}$$

$\mu\nu \neq 0$ denotes all ligands. Then the Sternheimer function $\gamma(r)$ can be replaced with γ_∞, with a value of -8.70 (Lauer et al., 1979)—in this context, γ_∞ is differently defined than in crystal field or point charge approximations.

In the principal axis system of the efg, the main component V_{zz} is just the sum of the above three contributions, whereas the valence part is given by

$$\Delta n_d = n_{x^2-y^2} + n_{xy} - n_{z^2} - (n_{xz} + n_{yz})/2$$
$$\Delta n_p = (n_x + n_y)/2 - n_z \tag{5.9}$$

These are the occupation numbers of the Fe-3d- and 4p-shell, respectively. Commonly, the anisotropy of the Fe(3d) shell is the major part for high-spin Fe^{2+}, whereas for high-spin Fe^{3+} the ligand contribution dominates.

A decisive task is now the construction of an appropriate model cluster—experience has shown that good results for Mössbauer parameters and magnetic properties (as desired in our case of synthetic fayalite) can be achieved if large negative cluster charges are avoided but, at the same time, the clusters must not be too small. However, this discrepancy may lead to severe convergency problems, since it is troublesome where and how to terminate a cluster. It is possible to add further polyhedra and to stabilize the, e.g., O(2p) orbitals, however, at the expense of CPU time. However, the effect of these polyhedra can be simulated by replacing third shell oxygens by fluorine atoms or inserting hydrogens. This measure will reduce the cluster charge and, at the same time, lead to reasonable orbital energies of the O(2p) orbitals. Model calculations on clusters with variable size have shown that size-converged results are obtained for oxygen atoms within a radius of 4.20 Å from the central iron atom (where the inserted fluorines and hydrogens preserved the charge neutrality of the whole cluster). Naturally, the clusters become comparably large—with convergency problems avoided, though—they comprise 363 and 325 valence orbitals for M1 and M2, respectively. The calculation of such large systems had been illusory during the pre-DFT times.

The DFT theory, in particular the special form derived by M. Grodzicki, is able to decompose the total quadrupole splittings into the various contributions mentioned above. The total value of the QS is in good agreement with the experimental results from SCMBS; It may be followed that the efg constitutes a sensitive criterion for the reliability of the calculated electronic structure, since in this case the sign and size of the efg is related to the split pattern of the t_{2g}-like levels. This is demonstrated by Table 5.3.

Table 5.3 Calculated quadrupole splitting (in mm/s), asymmetry parameter η and efg contributions for the M1 and M2 sites in synthetic fayalite

Site	ΔE_Q	η	val(3d)	val(4p)	coval.	ligand
M1	+3.126	0.72	+3.349	−0.034	−0.088	−0.101
M2	+3.208	0.36	+3.605	−0.113	−0.187	−0.097

It may be seen that the efg is dominated by the 3d-valence part related to the anisotropy of the Fe(3d) shell, and the other contributions remain between 1% and 5% of the leading term. Regardless of minor effects like spin-orbit coupling and temperature-dependent geometry variations, the calculated variation of the QS with temperature (modelled by thermal occupation of the t_{2g}-like orbitals) yielded rather good agreement with the T-dependence of the experimental QS (Fig. 5.9):

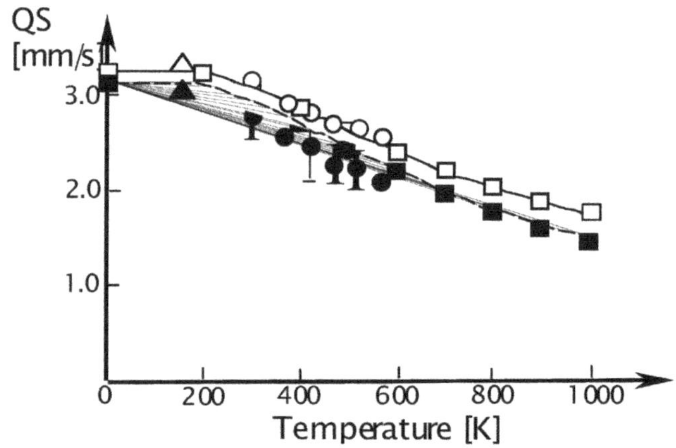

Figure 5.9 Temperature dependence of quadrupole splitting determined experimentally (circles), semiquantitatively (triangles) and full-quantitatively (squares). The closed and open symbols refer to M1 and M2 site, respectively, and the solid lines serve as eye guides.

It also turned out that the efg is, from the direction of its tensor components (eigenfunctions), severely sensitive to the slightest changes in the surrounding charge densities, especially on low-symmetry sites—small shifts can cause a complete overturn of the relevant efg. Therefore, we applied comparably large distances from the central ion on the M1 site (of 6 Å diameter) comprising a rather large cluster size, and smaller distances from the Fe^{2+} ion on the comparably high symmetric M2 site (4 Å diameter). In any case, the theoretical approach using SCC-Xα calculations has taught us that large cluster sizes are essential for the understanding of the fayalite problem, while the restriction to nearest-neighbour ions, as implied by the strong decrease of the potential, is insufficient.

These findings also correspond to the magnetic behaviour of fayalite at low temperatures. The comparably low-symmetrical M1 position displays a canted array of Fe^{2+} magnetic moments coupled through superexchange interaction to the collinearly ordered moments on the comparably high symmetrical M2 site. As the DEDs are strongly related to d-electron distributions, which are relevant for magnetism here, we may follow that the rather complicated magnetism on the M1 site is due to the spin frustration caused by the dominant influence of the M2 moments where the efg is rather stable and hence, the direction of the magnetic moments uniform (parallel to **b**). On this site, the magnetic moments keep this preferred orientation (SG Pnma) at all temperatures below the Néel point.

Concluding, the spin-polarized SCC-Xα calculations (fully quantitative method) show that with comparably large cluster sizes up to 97 atoms the efg becomes stable and good quantitative agreement between theory and experiment can then be achieved.

5.4 Establishing the Semi-Quantitative efg with the DEN (Difference Electron Nanoscope)

As already mentioned above, the basic data for the calculation of the semi-quantitative efg have been provided by the very accurate synchrotron diffraction study of Armin Kirfel (Kirfel and Lippmann, 2002). The experiments were performed at HASYLAB, Hamburg, Germany, using synchrotron radiation of 100.3 keV in order to obtain essentially absorption and extinction free data for the electron density modelling. The room temperature experiment was carried out using a synthetic fayalite single crystal sphere (grown by the Czochralski method) of 0.25 mm diameter on the Triple-Crystal Diffractometer BW5. Thousands of reflections were recorded arriving at a data set of 804 unique reflections with an internal agreement factor $R(F^2)$ = 0.0055 (!). This very low value was achieved with a refinement according to the independent atom (IAM) and generalized scattering factors models (GSF with multipoles up to order 7) with the VALRAY programme package (Stewart et al., 1996), the latter providing considerably better *R*-values.

The static deformation density plots in the publication of Kirfel and Lippmann (2002) give a nice example of nowadays 2D contour maps, the predecessors of our DED images discussed below. The difference from the latter, besides the lower dimensionality, is the usage of multipole refinement which may be considered as a very accurate, but *theoretical* reproduction of mainly 3d difference electron densities. The DEDs of our nanoscope, however, are produced by the difference of experimental and spherically refined structure factors, which creates *experimental* DEDs. In other words: If the experimental data were perfectly matched by multipole refinement, the DEDs would become zero.

The input data of our nanoscope, however, were created by long listings of observed and calculated (spherical symmetric) structure factors together with their relevant *hkl*'s and phase angles. According to the procedure mentioned above (cf. Section 4.3.2), by a modified FFT and the program EVOX a data file containing the fayalite DEDs has been produced. In the crystal-specific input of EVOX, however, it is eminently important to correctly choose the correct maximum boundaries of the desired cluster under consideration, i.e. with 6 Å size for the M1 position and 4 Å for M2.

This is a direct consequence of our previous DFT calculation where it has been evaluated how large the necessary cluster has to be chosen in order to obtain a stable efg. This is also a question of calculation economics, since the necessary memory space and CPU time increases with the cube of the distance between the central ion and the surrounding difference electron peaks. The same value of cube sizes, together with the lattice parameters, is demanded in the initial window of our DEN (difference electron nanoscope) program called DEDLOT. The images normally have to be created for each relevant crystallographic site separately, e.g., in the case of fayalite for the M1 position of the Fe^{2+} ion at 0.5, 0.5, 0.5 in the middle of the unit cell and for the M2 site of the other sort of Fe^{2+} ion at 0.22, 0.75, 0.48. It may happen, due to the size of the considered cluster, that, together with O^{2-} and Si^{4+} ions, the other Fe^{2+} ion might be visible at the boundary of the considered DEN image, but this is no disadvantage or even error source.

Like in the DFT calculation, we are working with clusters or, more accurately, with concentric coordination spheres around the given ion or nucleus. In order to provide a better visual orientation for the viewer, the position of the central nucleus (and for the other nuclei as well) can be displayed as green-coloured spheres which are created by a second input file with the calculated central positions of the different ions from crystallography.

Another, very important, topic is the so-called rendering size. As mentioned above (Section 4.3.4), below the initial black frame of the initial DEN window, a slider can be moved (one for the DEDs and one also for the nuclei if desired) to choose an appropriate size of the single hyperareas or DED clouds. For example, taking the initial render values, we will get a sort of "red Lasagne" —impossible to distinguish the different clouds. By moving the slider to a suitable value, which is provided by experience, the DED clouds converge to reasonable sizes, being not at all arbitrary (as the procedure might infer), but determined by the relevant size of the widely used 2D contour maxima and identical for different crystallographic sites.

The next step is the choice of the DEDs around the central ion on the chosen site in order to calculate an efg according to a point charge model. As mentioned above, the DEDs or hyperareas—which are in fact 4D because they express the various densities by different shells per cloud dependent on the three coordinates in space—are integrated. This is performed by moving the pointer to the relevant cloud and selecting it per mouse click. In order to indicate the successful selection, the relevant cloud is highlighted by changing the colour from red to yellow. This procedure is for large clusters rather arduous, considering the fact that we always should select symmetry equivalent clouds around the central ion and should not forget any. In order to avoid this, the above-mentioned tool of the "collective selection" of DED clouds within a given wire-framed sphere in dark-blue (with a diameter defined by the user) was introduced (Section 4.3.4). The program will calculate all DED clouds within this sphere and highlight them.

We did this for the M1 position of fayalite where we did not suffer from series termination errors. For the M2 position, however, it was necessary to find the artefacts by using our Fourier ghost buster program SELECT. By this it was possible to let out those clouds which were identified as being Fourier ghosts and to only select the real existing DEDs. The reward was—it should be anticipated at this place—that the evaluated efg well met the experimental one whereas the efg with all clouds included (also the artefacts) was erroneous to a large extent.

On the basis of the selected (highlighted) DED clouds, the relevant efg for each relevant site was calculated by just choosing the menu point "Calculate EFG". As mentioned above, the DEDs are integrated and the secular equation solved by application of the Jacobi algorithm, since a multitude of DED values in a certain arrangement around the central ion needs a diagonalization of the efg tensor to the main axes system. As described above, the eigenvalues and eigenvectors of the efg on the considered crystallographic site are given in a listing. These values may be easily compared with the associated ones of the experimental method from SCMBS and the DFT procedure—in the following section, a table and a stereogram with these data will be provided (Section 5.5).

Here, a three-dimensional perspective view of the DEDs together with a light-blue wire frame model of the efg's graphical representation is plotted in the main window of the DEN just by selecting the menu point "Draw EFG" and adjusting the relative size of the efg by the slider mentioned above (Section 4.3.4). By carefully applying the colours of the different "SLOTs", the perspective impression of the DED clouds floating in space can be enhanced. In the initial set up of the DEN, the DEDs are drawn in red, the highlighted ones in yellow. By slightly changing the render values and sophistically varying the colours of the SLOTs with the RGB-sliders (or else), and their opacity at the same time, the DED clouds will become more and more plastic in the sense of "three-dimensionally visible". The newly chosen colour slot determines the prevailing colour, subsequent slots can provide a nice "shadowing".

This part of the DEN application is not so much challenging the scientific abilities of the user but merely those of fine arts or design.

However, in order not to tantalize the reader too much, several DEN images of special views of the fayalite unit cell, the occurring DEDs and the relevant efg's are provided in the following.

Let us start with the M1 position at 0.5, 0.5, 0.5 (centre of the unit cell of synthetic fayalite) and different views along the {111} and {101} directions—looking under an angle of 45° from above towards the centre of the 6 Å cube mentioned above (Figs. 5.10–5.18).

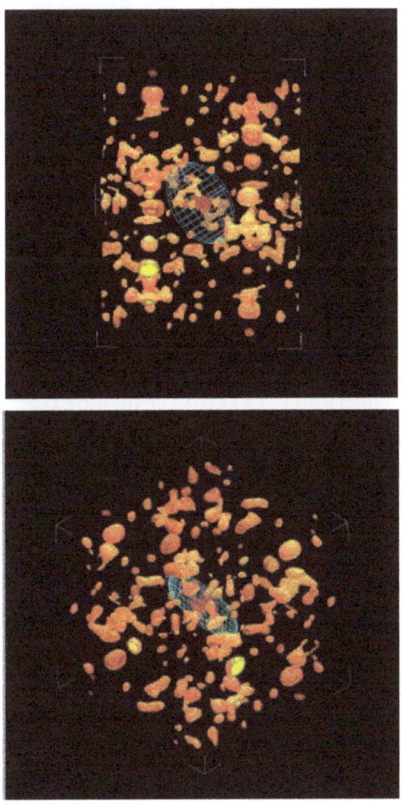

Figure 5.10, 5.11 (Above) 3D difference electron densities (DEDs) on the M1 site within a cube of 6 Å edge length and centre at 0.5, 0.5, 0.5 of a synthetic fayalite unit cell. The efg around the Fe^{2+} ion on this position is displayed as a light blue wire frame model. View along the [101] direction (upper image) and [111] (lower image), respectively.

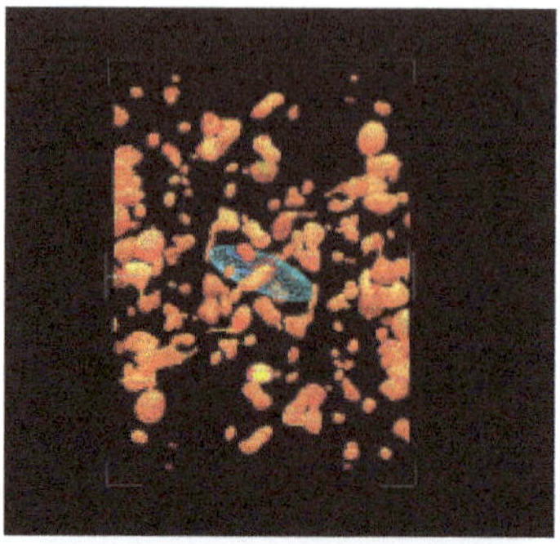

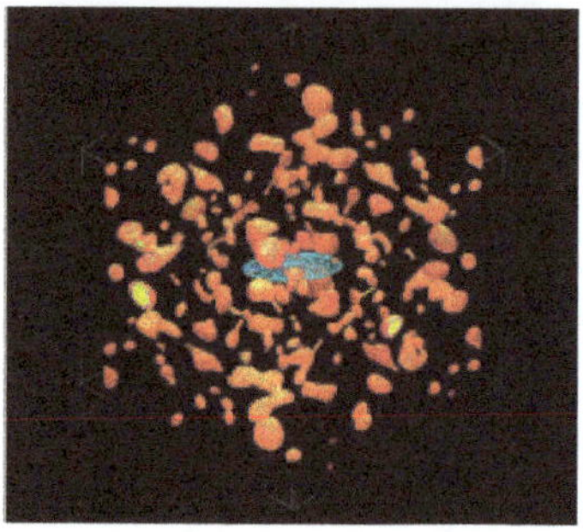

Figures 5.12, 5.13 (Above) 3D difference electron densities (DEDs) on the M1 site within a cube of 6 Å edge length and centre at 0.5, 0.5, 0.5 of a synthetic fayalite unit cell. The efg around the Fe^{2+} ion on this position is displayed as a light blue wire frame model. View along the [011] direction (upper image) and [-111] (lower image), respectively.

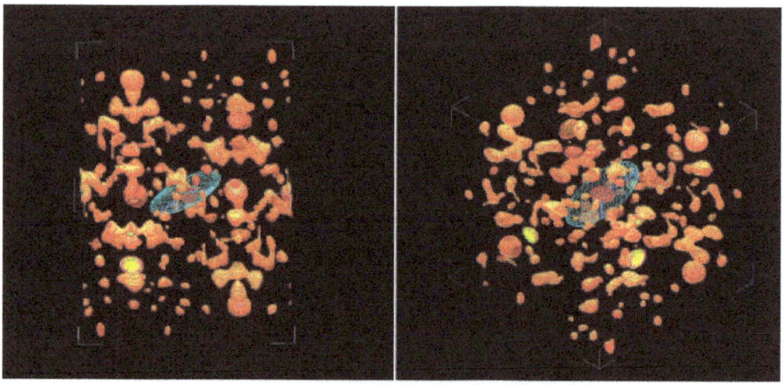

Figure 5.14, 5.15 (Right) 3D difference electron densities (DEDs) on the M1 site within a cube of 6 Å edge length and centre at 0.5, 0.5, 0.5 of a synthetic fayalite unit cell. The efg around the Fe^{2+} ion on this position is displayed as a light blue wire frame model. View along the [-101] direction (left image) and [-1-11] (right image), respectively.

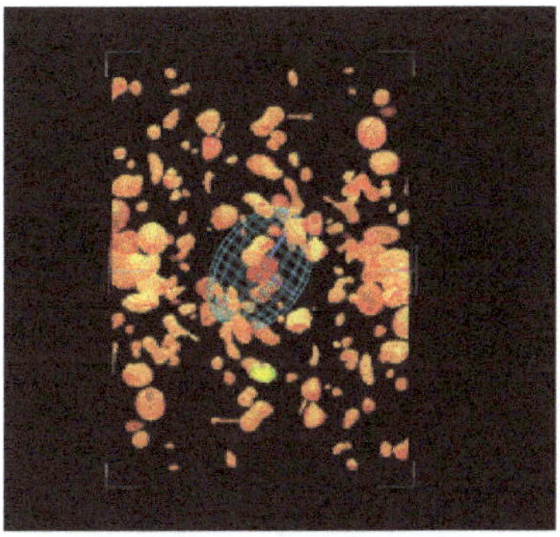

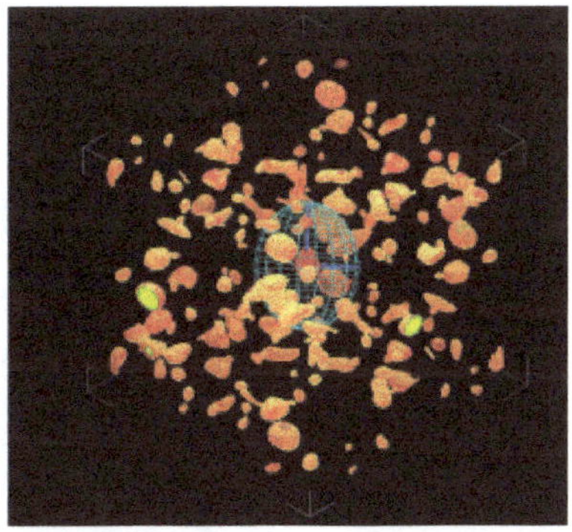

Figure 5.16, 5.17 (Above) 3D difference electron densities (DEDs) on the M1 site within a cube of 6 Å edge length and centre at 0.5, 0.5, 0.5 of a synthetic fayalite unit cell. The efg around the Fe^{2+} ion on this position is displayed as a light blue wire frame model. View along the [0–11] direction (left image) and [1–11] (right image), respectively.

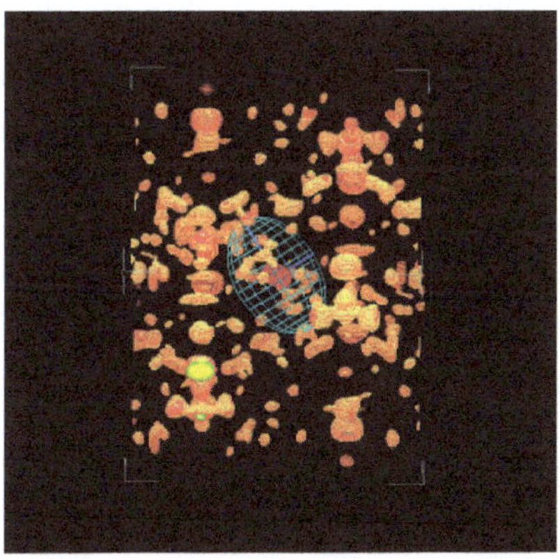

Figure 5.18 3D difference electron densities (DEDs) on the M1 site within a cube of 6 Å edge length and centre at 0.5, 0.5, 0.5 of a synthetic fayalite unit cell. The efg around the Fe^{2+} ion on this position is displayed as a light blue wire frame model. View along the [101] direction, like Fig. 5.10.

On these images, the efg tensor can be nicely seen from different perspectives. It is recommended to compare these images with the three-dimensional sketch of the unit cell provided by the crystallographic display routine CaRIne (Fig. 5.1). Since this may be difficult for the spatial sense of the reader, three views of the M1 position along the main axes *c* ([001] direction), *a* [100] and *b* [010] are given in the following together with an overlaid

CaRIne image showing the different nuclei (colour notation like in Fig. 5.1) together with the surrounding oxygen octahedra (Figs. 5.19–5.24):

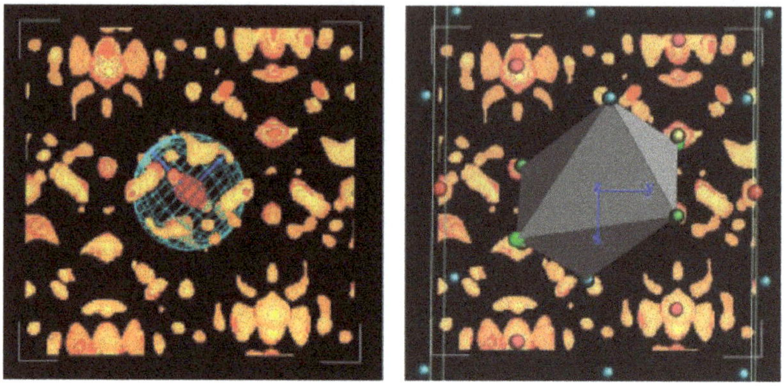

Figure 5.19, 5.20 3D difference electron densities (DEDs) on the M1 site within a cube of 6 Å edge length and centre at 0.5, 0.5, 0.5 of a synthetic fayalite unit cell. The efg around the Fe^{2+} ion on this position is displayed as a light blue wire frame model. View along the [001] direction (both images) and oxygen octahedron drawn in grey together with the relevant ions (right image, colour convention similar to Fig. 5.1) for better spatial imagination.

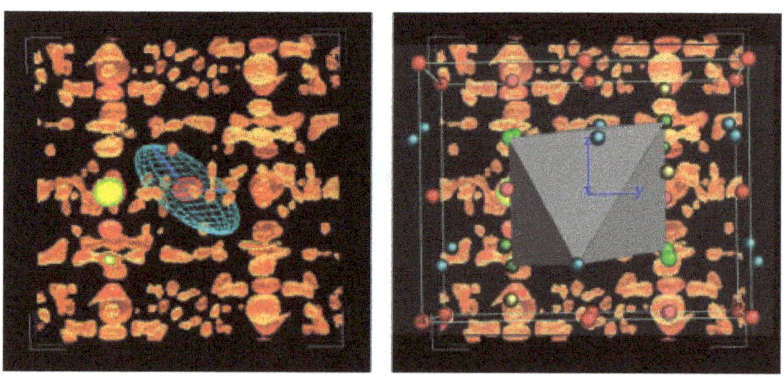

Figure 5.21, 5.22 3D difference electron densities (DEDs) on the M1 site within a cube of 6 Å edge length and centre at 0.5, 0.5, 0.5 of a synthetic fayalite unit cell. The efg around the Fe^{2+} ion on this position is displayed as a light blue wire frame model. View along the [100] direction (both images) and oxygen octahedron drawn in grey together with the relevant ions (right image, colour convention similar to Fig. 5.1) for better spatial imagination.

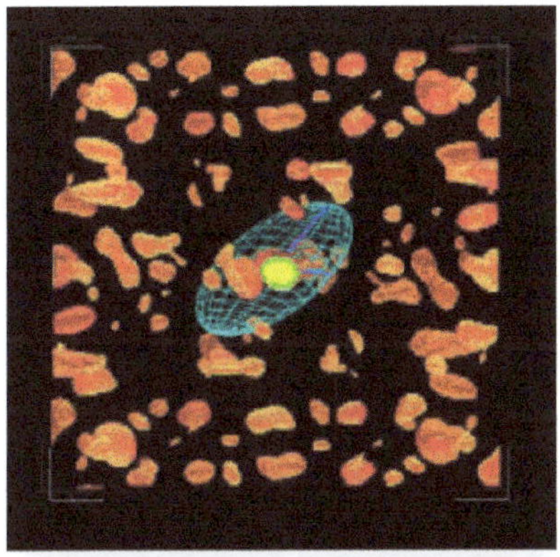

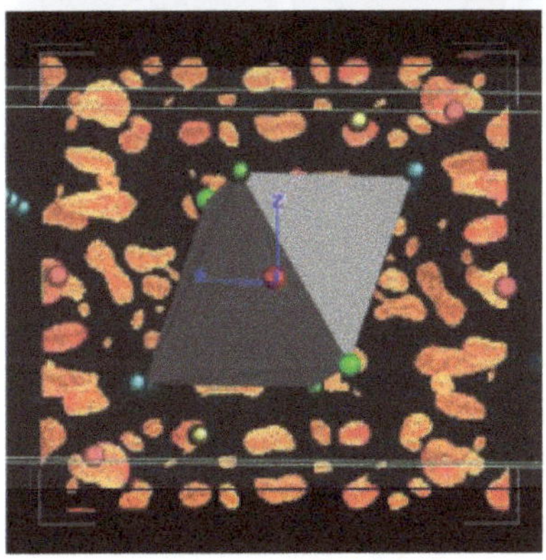

Figure 5.23, 5.24 3D difference electron densities (DEDs) on the M1 site within a cube of 6 Å edge length and centre at 0.5, 0.5, 0.5 of a synthetic fayalite unit cell. The efg around the Fe^{2+} ion on this position is displayed as a light blue wire frame model. View along the [010] direction (both images) and oxygen octahedron drawn in grey together with the relevant ions (lower image, colour convention similar to Fig. 5.1) for better spatial imagination.

From the previous DEN images, it may be seen how nicely the oxygen octahedron from different views correlates with the efg—it might be no 1:1 correlation, but this is by no means surprising:

According to the DFT results and our proceeding with the collection of DEDs of rather distant coordination spheres, the efg is not exactly aligned within the octahedron. However, these images merely serve as an illustration of the physical properties—the quantitative details may be read up from the efg table of the following section.

We also present here the features of the M2 position at 0.22, 0.75, 0.48 as again viewed along the [001], [100] and [010] direction, respectively (Figs. 5.25–5.30):

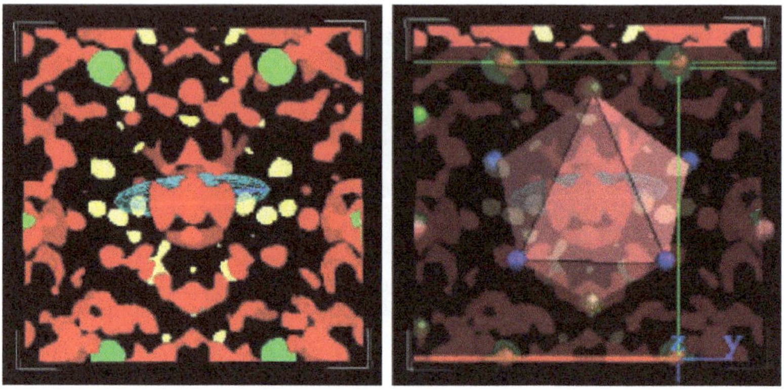

Figure 5.25, 5.26 3D difference electron densities (DEDs) on the M2 site within a cube of 4 Å edge length and centre at 0.22, 0.75, 0.48 of a synthetic fayalite unit cell. The efg around the Fe^{2+} ion on this position is displayed as a light blue wire frame model. View along the [001] direction (both images) and oxygen octahedron drawn in light red together with the relevant ions (right image, colour convention similar to Fig. 5.1) for better spatial imagination.

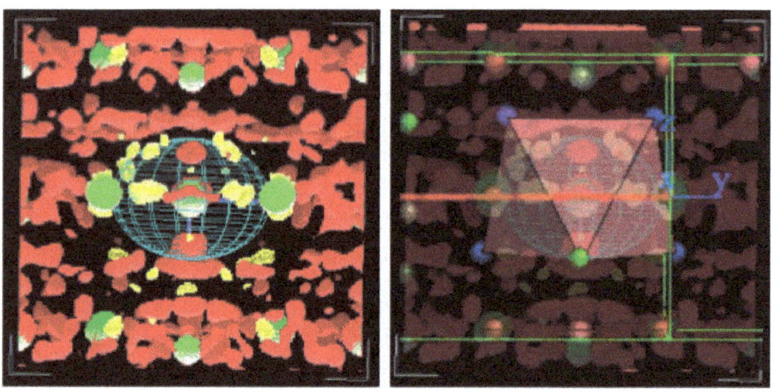

Figure 5.27, 5.28 3D difference electron densities (DEDs) on the M2 site within a cube of 4 Å edge length and centre at 0.22, 0.75, 0.48 of a synthetic fayalite unit cell. The efg around the Fe^{2+} ion on this position is displayed as a light blue wire frame model. View along the [100] direction (both images) and oxygen octahedron drawn in light red together with the relevant ions (right image, colour convention similar to Fig. 5.1) for better spatial imagination.

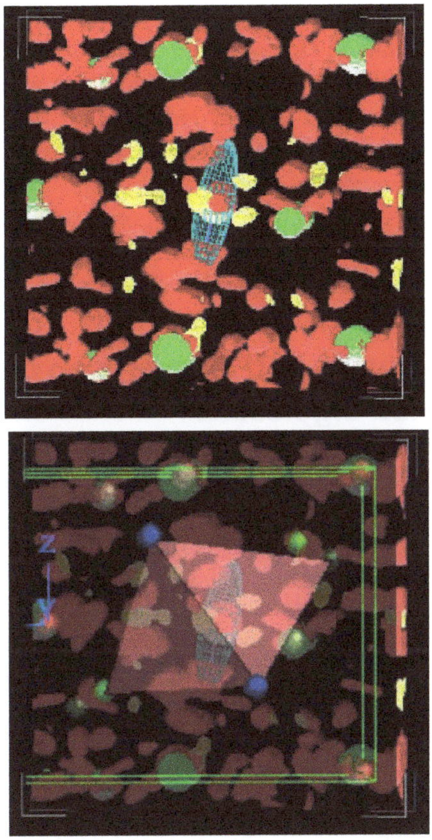

Figure 5.29, 5.30 3D difference electron densities (DEDs) on the M2 site within a cube of 4 Å edge length and centre at 0.22, 0.75, 0.48 of a synthetic fayalite unit cell. The efg around the Fe^{2+} ion on this position is displayed as a light blue wire frame model. View along the [010] direction (both images) and oxygen octahedron drawn in light red together with the relevant ions (lower image, colour convention similar to Fig. 5.1) for better spatial imagination.

The Fe^{2+} ions on the M2 position are situated on a comparably high symmetric site with local mirror planes. This can be directly seen at the surroundings and the oxygen octahedra, also denoted in Figs. 5.25–5.30, which show the rather eye-catching *m*-symmetry.

Correspondingly, the efg on this site, again displayed as a light-blue wire frame model, shows this rather high symmetry. Interestingly, the red colour dominates on these images—this is the consequence of the application of our Fourier ghostbuster program SELEC, by which several DED maxima could be identified as artefacts. This was successively done by enhancing the value of sineθ/λ and thus detecting the relevant peaks as being series termination errors. Hence, they were not selected as contribution to the calculation of the efg on M2—this has later been confirmed indirectly by the fact that the efg from this "corrected" distribution is in good agreement with the experimental one from SCMBS. Hence, we are confident in the correct application of our selection procedure.

5.5 Combination and Comparison of the Obtained efg Results

A rather clear method to compare the goodness of the different efg evaluations is a common table, where the angle values of the main efg axes (eigenvectors) with respect to the crystallographic axes are listed (Tables 5.4 and 5.5). First, we quote the experimental values from SCMBS measurements that are the reference point for the DFT and DEN evaluations. The former is given as the second column/row in Tables 5.4 and 5.5, the latter as the third one.

Table 5.4 Results of our efg determination on the M1 site (synthetic fayalite) yielding all angles of the efg components with respect to the crystallographic axes (SG Pnma)

efg-comp.	Exp	Full quant Grodzicki	Semi-DEN
M1			
V_{xx},**a**	48.4	61.7	46.9
V_{xx},**b**	63.5	51.2	62.8
V_{xx},**c**	55.5	51.9	55.3
V_{yy},**a**	123.8	130.8	123.8
V_{yy},**b**	39.5	43.8	33.8
V_{yy},**c**	88.9	103.3	90.1
V_{zz},**a**	120.0	126.2	118.3
V_{zz},**b**	117.0	107.1	108.4
V_{zz},**c**	34.5	41.2	34.7

Note: First column: experimental, according to SCMBS. Second column: Full-quantitatively, according to DFT method/M. Grodzicki. Third column: Semi-quantitatively, according to Difference Electron Nanoscope (DEN).

Table 5.5 Results of our efg determination on the M2 site (synthetic fayalite) yielding all angles of the efg components with respect to the crystallographic axes (SG Pnma)

efg-comp.	Exp	Full quant. Grodzicki	Semi-quant. DEN
M2			
V_{xx},**a**	178.9	166.7	171.0
V_{xx},**b**	90.0	90.0	98.1
V_{xx},**c**	90.0	76.7	94.0
V_{yy},**a**	89.0	76.7	85.9
V_{yy},**b**	113.0	90.0	90.3
V_{yy},**c**	-16.5	13.3	175.9
V_{zz},**a**	89.9	90.0	98.1
V_{zz},**b**	23.0	0.0	8.1
V_{zz},**c**	73.5	90.0	90.9

Note: First column: experimental, according to SCMBS. Second column: Full-quantitatively, according to DFT method/M. Grodzicki. Third column: Semi-quantitatively, according to Difference Electron Nanoscope (DEN).

Interestingly, the error of the DFT values for the angles of the efg seems to be larger than the one of the DEN results. This is not surprising, since the DFT values are obtained from highly complex quantum mechanical calculations with no adjustable parameters and almost ideal orbitals like those mentioned in textbooks. Insofar the evaluation of these orbitals is close to the multipole refinement of Armin Kirfel, where also superpositions of orbitals have been used.

The DEN method, however, collects the DEDs (merely stemming from *observed* structure factors) with all details—also with those which are not met by the "ideal" orbitals of the DFT approach. Insofar, the DEN method provide "better" results. This may be visible in Fig. 5.31, where stereograms of the different efg axes with respect to the crystallographic axes are drawn.

Indeed, the poles of the DEN efg for M1 and M2 lie closer to those of the experimental ones than the efg poles of the DFT. However, it would be erroneous to believe that we could abdicate DFT calculations at all. These have given us valuable information concerning details of efg evaluations, e.g., the necessary cluster size, the overturn of the efg in case of too small clusters a.s.o.

Hence, only the synergy of the different methods might lead to consistent and reliable results.

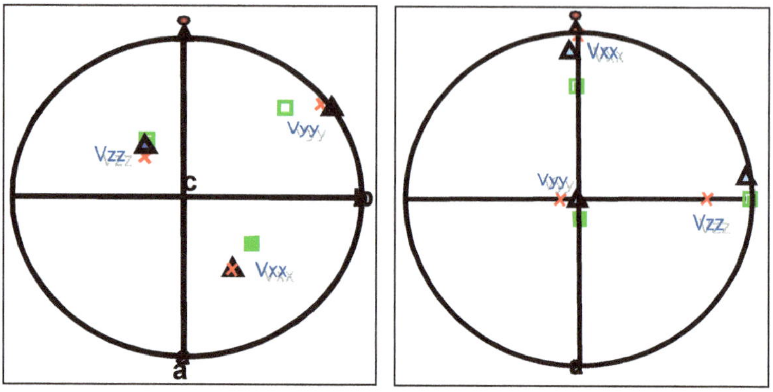

Figure 5.31 Stereographic projections of the poles of the efg components with respect to the crystallographic axes of synthetic fayalite on M1 (left diagram) and on M2 (right one). The experimental points are given as red crosses, the DFT calculations as green squares and the DEN points as blue triangles.

An objection sometimes raised against our DEN procedure is that "we might select DEDs until the desired (experimental) efg is achieved". This is absolutely untrue: Besides the neglecting of obvious artefacts we do not select single DEDs, only those of whole clusters like the ones of the DFT method. Moreover, it is completely improbable to select DEDs so sophistically and purposefully that we finally arrive at nine correct efg components for each crystallographic site—we are no Einsteins. Hence, it is not possible to "construct" an efg just by manipulation of input data.

5.6 Derivation of the Magnetic Structure by Neutron Diffraction

As already mentioned above, the magnetic structure of synthetic fayalite was first evaluated by neutron diffraction, as well as an overwhelming multitude of other examples. Therefore, this method has become commonly accepted as the most powerful mean to derive a magnetic structure in the solid state. However, SCMBS can—in favourite cases—provide similar results, always with the confinement that only limited elements/nuclides are accessible. Insofar neutron diffractometry will always keep its outstanding position within the research of magnetism, since even two submethods can be used to derive a magnetic structure:

The integrated neutrons, as described in detail above, are able to let determine the magnetic moment and its components on a given crystallographic site, whereas polarized neutrons can provide maps of the distribution of magnetic moments within the unit cell of the given sample ("spin density maps"). These resemble classical contour maps, with the difference that—contrary to X-ray diffractometry—not electrons in general determine these maps, but only "magnetic" electrons resp. moments.

Unfortunately, due to not previsible effects like antiphase domains with a distribution of close to 50:50 (which are "lethal" for the evaluation of magnetic structure factors, cf. Section 3.3.4.7), in the case of synthetic fayalite an experiment with polarized neutrons has failed. However, integrated neutron diffractometry was successful, so that Table 3.2 for temperatures below the Néel point could be established (Müller, 1985):

It turned out that the magnetic moments on the M2 position remained rather stable (within the "normal" Brillouin decrease with increasing temperature) and were aligned parallel to **b** (SG Pnma).

The M1 magnetic moments, however, displayed an increasing canting with decreasing temperatures below T_N—hints for a supposed second magnetic transition around $T \approx 23$ K could not be confirmed. Temperature-dependent recordings of several reflections characteristic for canting could not solve this problem as well.

In Fig 5.2, the canting mentioned above is displayed visually as a perspective view of the unit cell.

5.7 The Internal Magnetic Field *H*(0) Determined by SCMBS

In our publication from 1996 (Lottermoser et al., 1996), the possibility to derive even cantings of magnetic moments by SCMBS was pointed out, since there is a strong correlation between the strength of the internal magnetic field as derived by SCMBS and the magnetic moment from neutron diffractometry or susceptibility measurements. These fields (comprising 32 and 15 T for synthetic fayalite at low temperatures, respectively) are strong enough to align the respective moments, although there may be concluded no 1:1 correlation between both. Since by SCMBS the orientational angles of $H(0)$ with respect to the axes of the efg and furthermore, the latter with respect to the crystallographic axes, can be determined, it is possible to derive canted magnetic structures from SCMBS measurements. This has even been demonstrated in the case of $LiFeSi_2O_6$, where our (correct) prediction of a canted structure was even earlier than the subsequent confirmation by neutron diffractometry (Baum et al., 1988). In the two stereograms shown in Fig. 5.32, the poles of the internal magnetic field $H(0)$ and efg are displayed for three temperatures below T_N for synthetic fayalite (M1 and M2).

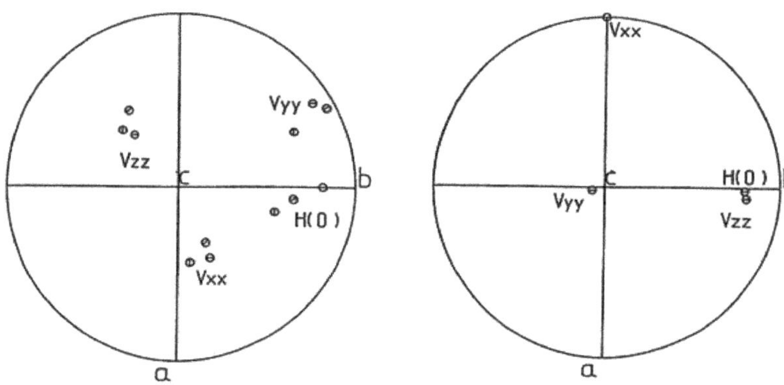

Figure 5.32 Stereographic projections of the poles of the efg components with respect to the crystallographic axes of synthetic fayalite on M1 (left diagram) and on M2 (right one). The experimental points from SCMBS at low temperatures below T_N are given as open circles whose inlet bars denote by their alignment the grade of low temperature (the more vertical the lower in temperature).

These findings were in full agreement with the neutron diffractometry results. However, this was not all:

By careful inspection of the temperature dependence of $H(0)$ and the corresponding magnetic moments on M1, it turned out that neutron diffractometry has "planished" important details of this T-dependence around the supposed second transition point. Retrospectively, this is not surprising: The time for neutron diffraction measurements is confined to several days and this time frame is commonly used for some few magnetic reflection measurements dependent on T and some recordings of also few data sets below T_N.

In this case now, the details of the dependency non-surprisingly escaped the neutron researcher: By SCMBS we could identify the supposed second transition as a collective discontinuous "jump" of the magnetic moments from one canted structure above $T_K \approx 20$ K to another one below. This behaviour is due to the high frustration grade of the M1 spins and may be compared with a "lock in place" thus gaining a new minimum of the moments' free energy.

Besides these details, the canting properties can be nicely shown with DEN images, where the vector of the internal magnetic field from SCMBS has been inserted into the unit cell for two temperatures (50 K and 10 K) below T_N and for different views along [001], [100] and [010], respectively (Figs. 5.33–5.35).

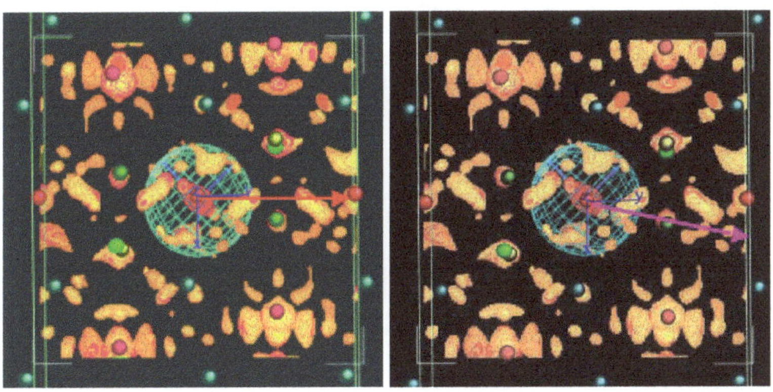

Figure 5.33 3D difference electron densities (DEDs) on the M1 site within a cube of 6 Å edge length and centre at 0.5, 0.5, 0.5 of a synthetic fayalite unit cell. The efg around the Fe^{2+} ion on this position is displayed as a light blue wire frame model. View along the [001] direction (both images) together with the vector of the internal magnetic field from SCMBS or of the magnetic moment from neutron diffraction, respectively. T = 50 K (collinear AF, red arrow, left image) and T = 10 K (canted AF structure, pink arrow, right image).

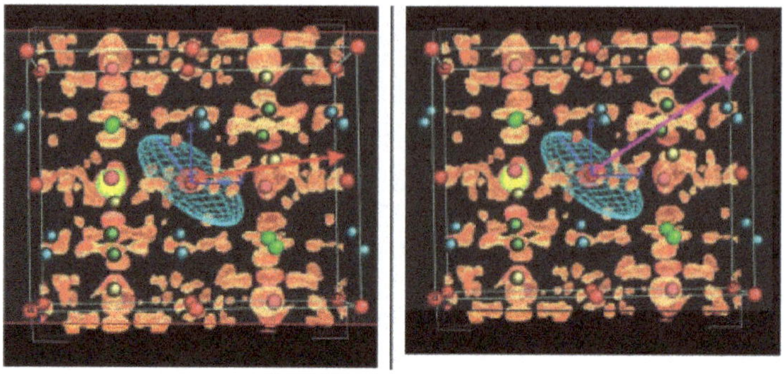

Figure 5.34 3D difference electron densities (DEDs) on the M1 site within a cube of 6 Å edge length and centre at 0.5, 0.5, 0.5 of a synthetic fayalite unit cell. The efg around the Fe^{2+} ion on this position is displayed as a light blue wire frame model. View along the [100] direction (both images) together with the vector of the internal magnetic field from SCMBS or of the magnetic moment from neutron diffraction, respectively. T = 50 K (collinear AF, red arrow, left image) and T = 10 K (canted AF structure, pink arrow, right image).

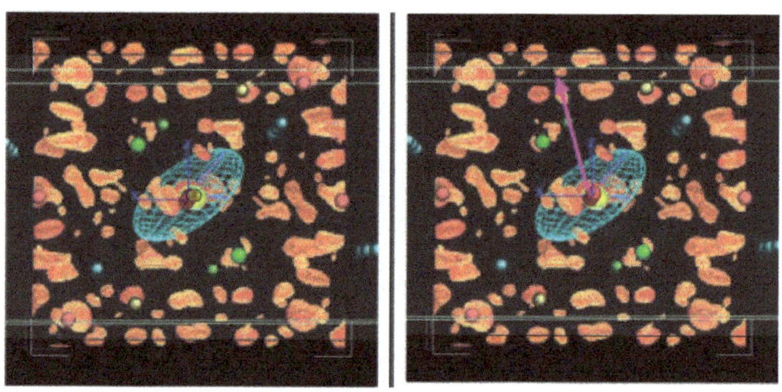

Figure 5.35 3D difference electron densities (DEDs) on the M1 site within a cube of 6 Å edge length and centre at 0.5, 0.5, 0.5 of a synthetic fayalite unit cell. The efg around the Fe^{2+} ion on this position is displayed as a light blue wire frame model. View along the [010] direction (both images) together with the vector of the internal magnetic field from SCMBS or of the magnetic moment from neutron diffraction, respectively. T = 50 K (collinear AF, left image) and T = 10 K (canted AF structure, pink arrow, right image). On the left image we see the vector of $H(0)$ pointing towards the viewer (yellow peak).

The same images were generated for the M2 position (Figs. 5.36–5.38), but the latter are not so striking because of the rather simple spin structure with moments parallel to **b**:

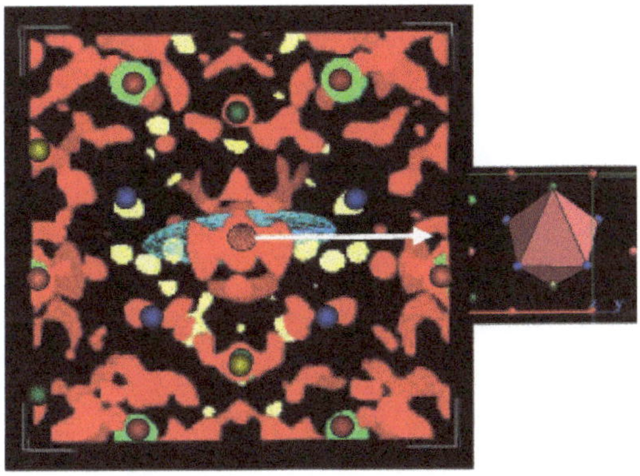

Figure 5.36 3D difference electron densities (DEDs) on the M2 site within a cube of 4 Å edge length and centre at 0.22, 0.75, 0.48 of a synthetic fayalite unit cell. The efg around the Fe^{2+} ion on this position is displayed as a light blue wire frame model. View along the [010] direction (both images) and oxygen octahedron drawn in light red together with the relevant ions (smaller image on the right, colour convention similar to Fig. 5.1) The vector of the internal magnetic field from SCMBS or of the magnetic moment from neutron diffraction, respectively, is drawn into the left image as a white arrow. T = 50 K (collinear AF, prevailing on M2).

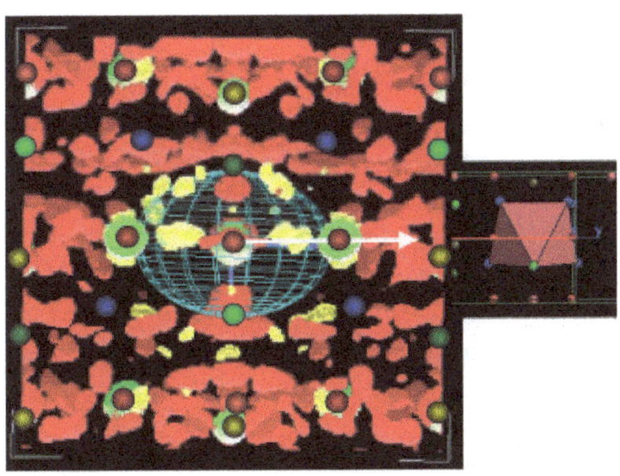

Figure 5.37 3D difference electron densities (DEDs) on the M2 site within a cube of 4 Å edge length and centre at 0.22, 0.75, 0.48 of a synthetic fayalite unit cell. The efg around the Fe^{2+} ion on this position is displayed as a light blue wire frame model. View along the [100] direction (both images) and oxygen octahedron drawn in light red together with the relevant ions (smaller image on the right, colour convention similar to Fig. 5.1). The vector of the internal magnetic field from SCMBS or of the magnetic moment from neutron diffraction, respectively, is drawn into the left image as a white arrow. T = 50 K (collinear AF, prevailing on M2).

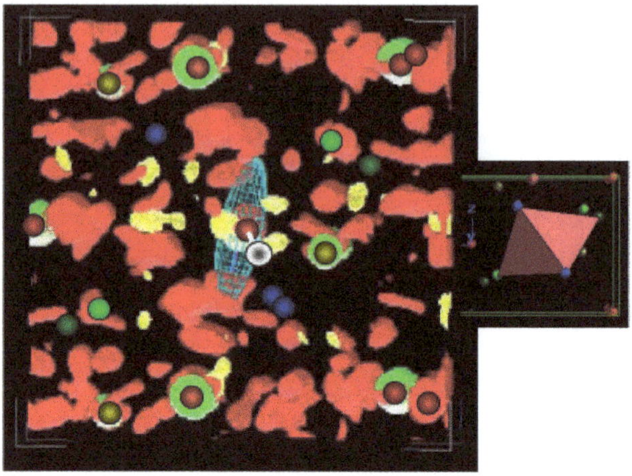

Figure 5.38 3D difference electron densities (DEDs) on the M2 site within a cube of 4 Å edge length and centre at 0.22, 0.75, 0.48 of a synthetic fayalite unit cell. The efg around the Fe^{2+} ion on this position is displayed as a light blue wire frame model. View along the [010] direction (both images) and oxygen octahedron drawn in light red together with the relevant ions (smaller image on the right, colour convention similar to Fig. 5.1). The vector of the internal magnetic field from SCMBS or of the magnetic moment from neutron diffraction, respectively, is drawn into the left image as a white arrow pointing towards the viewer. T = 50 K (collinear AF, prevailing on M2).

Since the magnetic moments, at least with respect to their orientation, may be identified with the direction of the internal magnetic fields, we gain a rather consistent view of the magnetic properties in the fayalite unit cell.

5.8 The Contribution of the DEN to Magnetic Coupling

The properties of magnetic coupling are a very inspiring topic within solid-state physics. From the first alleged detection by a Greek shepherd in antique history, over the applications of "magnetic stones" as floating in water containers forming early compasses and the surprising finding of the Bohr-van Leeuwen theorem (that a classical dipolar interaction is not sufficient to explain collective magnetism), the phenomenon of magnetism and magnetic coupling has always been something "magic". Nowadays, we know that these interactions are caused by Heisenberg's direct exchange or, if this explanation is not applicable, by indirect or superexchange or even more sophisticated theories (cf. Lottermoser, 2015).

You may stunningly observe how the metallic ballpen of an incautious visitor of your high-field lab is torn out of his suit's front pocket, traverses the space like a projectile, crashes through several layers of stainless steel, nitrogen and helium and ends up in the central tube of a 10 T superconducting magnet thus causing a terrible mess of a burst cryostat, steam clouds and a quenched magnet—a damage of thousands of Euro. You may then calculate the forces and energies, applying what you have learnt about inhomogeneous magnetic fields acting even in the empty space and developing destructive energies like a big firearm (and more), but you will never really *understand* what has happened.

Contrary to our everyday experiences of cause and effect of forces, the magnetic interactions in particular remain dark and—as such—a source of fascination.

Since in the present case of synthetic fayalite, direct exchange as the cause of the amazing magnetic interactions at low temperatures can be ruled out—due to the comparably large distances between the Fe^{2+} ions, only superexchange remains to be considered as the source of magnetic coupling here. This kind of interaction is caused by pairs of magnetic ions (in our case 3d cations) and intermediate oxygen anions with 2p orbitals. The latter may be polarized—under certain geometrical conditions—and leading to a "transferred" coupling to the next nearest neighbour 3d cation. This transferred interaction can be ferromagnetic or antiferromagnetic; in our case of synthetic fayalite, the Fe^{2+} ions couple antiferromagnetically. As has been shown by Bonnet et al. (1979) in the case of YIG and Lottermoser and Fuess (1992) in the case of Mn_2SiO_4, the relevant oxygens as being responsible for the transferred coupling should bear a comparably small amount of magnetic moment. This can easily be verified by polarized neutron diffractometry, but also has been found in a detailed X-ray study (Fujino et al., 1981). Since in the case of fayalite, we unfortunately do not dispose of polarized neutron results, we have to rely on the latter investigation, where a non-vanishing electronic density—which has to be identified as a magnetic electron density due to the applied method—has been detected on the oxygen O(2). The same was true, this time with real polarized neutron diffractometry for the isostructural Mn_2SiO_4 (Lottermoser and Fuess, 1992).

Now we may consider a very significant DEN image of the M1 cations. In the close neighbourhood of them, we find the bridging oxygens O(2) together with electronic structures resembling 2p orbitals (Fig. 5.39)!

Figure 5.39 3D difference electron densities (DEDs) on the M1 site within a cube of 6 Å edge length and centre at 0.5, 0.5, 0.5 of a synthetic fayalite unit cell. The efg around the Fe^{2+} ion on this position is displayed as a light blue wire frame model. View along the [001] direction with neighbouring Fe1 and Fe2 ions on M1 and M2, respectively. The shape of a Si-tetrahedron is drawn in yellow, the central oxygen octahedron together with the relevant O^{2-} ions as green lines in order to generate a spatial impression. White lines are drawn in between the central iron ion on 0.5, 0.5, 0.5 and its next iron neighbours in order to mark a common plane. Along certain superexchange pathways close to special oxygen ions, difference electron densities (here: most probably residual magnetic moments) can be recognized, as the theory of magnetism in special solid-state materials has predicted.

This may provide strong evidence for the existence of superexchange coupling, since such eye-catching details are missing in the neighbourhood of, e.g., O3 anions. The coupling from Fe^{2+} on M1 to Fe^{2+} on M2, however, may be provided by the slight DED intensity close to O(1). However, we should avoid overinterpretations: The Fe^{2+} cations on M2 together with their DEDs severely suffer from series-termination errors—the highly symmetrical DEDs in the close neighbourhood of the M2 nuclei are with high probability just artefacts.

Chapter 6

Summary and Outlook

Without exaggeration, it may be stated that the present book represents my life's work. As already mentioned in a preceding chapter, I have not been eager to collect many experimental details in the hope to find a common thread: I wanted to "see the rainbow" in the waterdrop—therefore I mainly confined myself to the waterdrop "fayalite", examined it quite thoroughly by applying different methods and—hoping not to boost too much—found the rainbow. This should not be understood as "the" rainbow, but as another nice "stone at the beach of the Ocean of the scientific Truth" like Isaac Newton has formulated it so aptly.

The rainbow or nice stone is the perception that findings of diffractometry and spectroscopy (merely Mössbauer spectroscopy) can be unified via the common quantity of the electric field gradient.

The drive to unification is a mainspring in natural sciences; perhaps the most famous example is the search for the Unified Field Theory by Einstein and Heisenberg, and this impulse is ongoing up to the present days.

However, the diffractometry is represented by methods like neutron, X-ray or synchrotron diffraction—especially the latter warrants the large data sets necessary to establish a multitude of well-measured structure factors. The DEN is dependent on these—the very ostensive unit cell images of difference electron

The Difference Electron Nanoscope: Methods and Applications
Werner Lottermoser

ISBN 978-981-4774-01-7 (Hardcover), 978-1-315-19664-0 (eBook)
www.panstanford.com

densities put aside—the evaluation of a semi-quantitative efg is only possible with excellent data sets. On the other hand, Mössbauer spectroscopy serves to provide an experimental efg, however, confined to mostly iron bearing samples; NQR/NMR methods may extend the limit of available elements or nuclides, but also these are confined. The DFT, however, is—like diffractometry—not intrinsically restricted. Nowadays it is on such a high standard that efg's from not too complicated structures may be considered as being generally close to experimental ones. In this sense, the DEN with respect to DFT efg's may be considered as a rather universal tool. Besides the evaluation of efg's on different crystallographic sites within the unit cell, amazing images of 3D/4D difference electron density distributions are possible. This is not only question of aesthetics or delighting the viewer. Contrary to DFT molecular orbital plots or multipole refinement, these are no theoretical images or even simulations: They are representations of physical reality. The efg is *really* floating within the DEDs surrounding the selected ion—in this sense a comparably abstract quantity gains visual clarity. The distributions of DEDs represent *real* molecular orbitals—sometimes not easy to recognize, though—and not idealized hybrid orbitals as shown in textbooks. Therefore, we might gain new insights in structure-property relationships by regarding these images—the detection of superexchange oxygens and the relevant DEDs is one example. Besides magnetism, other physical properties like, e.g., optics/dichroism might benefit from the visual insights provided by the DEN.

However, a certain shortcoming still is the intrinsic reduction of the 3D images on a 2D computer screen. This issue can be compensated by a visual demonstration I have given during a talk: Different views of the efg and the DEDs towards the common centre (the crystallographic site under consideration) can be summarized to a movie. Unfortunately, it is not possible to show it in a static book, but this movie has provided a nice spatial impression of the contents of the unit cell just by the observation that details being closer to the viewer are moving faster than those in the background. By this measure, it could clearly be seen, which DED is in the foreground and which is in the background—even with the necessary graduations.

For the near future, artificial polarized or red/green double images shall be created in order to provide a vivid 3D impression of the unit cell.

Another aim is the establishment of a "real" nanoscope in the sense of a device: Extrapolating the achievements in computer and software performance during the last decades, it should be possible to create the different images "online": The user will put his/her sample into the experimental device of a neutron or synchrotron facility and will receive the views of the unit cell in almost "real time". We are still far away from this vision, also due to the necessary measuring time of the different reflections, but who knows?

Image plates/detectors for a simultaneous recording of reflections are already in operation and the computer power is still increasing rapidly. Perhaps, in a couple of years, the researcher will sit in the reactor or synchrotron facility and regard the unit cell of his sample by a binocular eyepiece like he nowadays enjoys the tunnel microscopic or HRTEM images.

References

Amthauer G, Pavicevic MK (eds) (2001) *Physikalisch-chemische Untersuchungsmethoden in den Geowissenschaften*, vol. 2, E. Schweizerbart'sche Verlagsbuchhandlung (Nägele und Obermiller), Stuttgart.

Bacon GE (1975) *Neutron Diffraction*, 3rd edition, Clarendon Press, Oxford.

Ballet O, Fuess H, Wacker K, Untersteller E, Treutmann W, Hellner E, Hosoya S (1989) Magnetisation measurements of synthetic olivine single crystals A_2SiO_4 with A = Mn, Fe or Co, *J Phys Condens Matter,* **1**, 4955–4970.

Barb D (1980) *Grundlagen und Anwendungen der Mössbauerspektroskopie* (Meisel W, ed), Akademie Verlag, Berlin.

Baum E, Treutmann W, Behruzi M, Lottermoser W, Amthauer G (1988) Structural and magnetic properties of the clinopyroxenes $NaFeSi_2O_6$ and $LiFeSi_2O_6$, *Z Krist,* **183**, 273–284.

Becker P, Coppens P (1974) Extinction within the limit of validity of the Darwin transfer equations, I General formalism for primary and secondary extinction and their applications in spherical crystals, *Acta Cryst,* **A30**, 148–153.

Blaha P (2010) Calculations of Mössbauer parameters in solids by DFT bandstructure calculations, *J Phys Conference Ser*, 217, 012009.

Blaha P, Schwarz K, Madsen G, Kvasnicka D, Luitz J (2014) WIEN2k, www.wien2k.at, Inst. f. Materials Chemistry, Technical University of Vienna, Vienna.

Bonnet M, Delapalme A, Fuess H, Becker P (1979) Polarized neutron study of covalency effects in yttrium iron garnet, *Phys Chem Sol*, **40**, 863–876.

Borchardt-Ott W (1995) *Crystallography*, 2nd edition, Springer, Berlin

Boudias C, Monceau D (1989–1998) CaRIne crystallography 3-1: The crystallographic software for research and teaching, Senlis, France.

Cooley JW, Tukey JW (1965) An algorithm for the machine calculation of complex Fourier series, *Math Comput*, **19**, 297–301.

DiBenedetto F, Bernardini GP, Borrini D, Lottermoser W, Tippelt G (2005) ^{57}Fe- and ^{119}Sn-Mössbauer study on stannite (Cu_2FeSnS_4)-kesterite (Cu_2ZnSnS_4) solid solution, *Phys. Chem. Minerals*, **31**, 683–690.

Ehrenberg H, Fuess H (1993) Analytical interpretation and simulation of the static magnetic properties of synthetic α-Fe_2SiO_4, *J Phys Condens Matter*, **5**, 3663–3672.

Fuess H (2001) Synchrotronstrahlung, in *Physikalisch-chemische Untersuchungsmethoden in den Geowissenschaften*, vol. 2 (Amthauer G, Pavicevic MK, eds), E. Schweizerbart'sche Verlagsbuchhandlung (Nägele und Obermiller), Stuttgart.

Fuess H, Ballet O, Lottermoser W (1988) Magnetic phase transition in olivines M_2SiO_4 (M = Mn, Fe, Co, Fe_xMn_{1-x}), in *Structural and Magnetic Phase Transitions in Minerals* (Ghose S, Coey JMD, Salje E, eds), Springer, Heidelberg, pp. 185–207.

Fujino K, Sasaki S, Takéuchi Y, Sadanaga R (1981) X-ray determination of electron distributions in forsterite, fayalite and tephroite, *Acta Cryst*, **B37**, 513–518.

Geiger CA, Armbruster T, Lager GA, Jiang K, Lottermoser W, Amthauer G (1992) A combined temperature dependent ^{57}Fe Mössbauer and single crystal X-ray diffraction study of synthetic almandine: evidence for the Gol'danskii-Karyagin effect, *Phys Chem Minerals*, **19**(2), 121–126.

Goldanskii VI (1963) *The Mössbauer Effect and its Applications in Chemistry*. Atomic Energy Review (Austria) 1.4.

Goldanskii VI, Makarov EF, Khrapov VV (1963) On the difference in two peaks of quadrupole splitting in Mössbauer spectra, *Phys Lett,* **3**(7), 344–346.

Gonser U (1975) From a strange effect to Mössbauer spectroscopy in *Mössbauer Spectroscopy* (Gonser U, ed), Springer-Verlag, Berlin-Heidelberg-New York.

Grant RW (1975) Mössbauer spectroscopy in magnetism: Characterization of magnetically ordered compounds, in *Topics in Applied Physics* (Gonser U, ed), Springer-Verlag, Berlin-Heidelberg-New York, vol. 5, p. 104ff.

Grodzicki M (1980) A self-consistent-charge Xα method, *J Phys B,* **13**, 2683–2692.

Grodzicki M (1985) Theorie und Anwendungen der Self-Consistent-Charge Xα Methode, Thesis of habilitation, Hamburg.

Grodzicki M, Männing V, Trautwein AX, Friedt JM (1987) Calibration of isomer shifts and quadrupole coupling constants for ^{119}Sn, ^{127}I and ^{129}I as derived from SCC-Xα calculations and Mössbauer measurements, *J. Phys,* (B)20, 5595–5625.

Grytsiv A, Rogl P, Berger St, Paul Ch, Michor H, Bauer E, Hilscher G, Godart C, Knoll P, Musso M, Lottermoser W, Saccone A, Ferro R, Roisnel T, Noel H (2002) A novel skutterudite phase in the Ni-Sb-Sn system—phase equilibria and physical properties, *J Phys Condensed Matter,* **14**(29), 7071–7090.

Halpern D, Johnson MH (1939) On the magnetic scattering of neutrons, *Phys Rev,* **55**, 898.

Haussühl S (1977) Kristallgeometrie, in *taschentext 64*, Physik-Verlag, Weinheim, p. 8ff.

Hock R (1990) Bestimmung der magnetischen Strukturen der Seltenen-Erd-Eisen-Granate des Erbiums und Terbiums durch elastische Neutronenstreuung, Thesis, University of Frankfurt, Frankfurt/M.

Hock R, Fuess H, Vogt T, Bonnet M (1991) Low temperature magnetic structure of erbium iron garnet, *Z Phys B Condensed Matter,* **82**, 283–294.

Karyagin SV (1963) *Dokl Akad Nauk SSSR,* **148**, 1102.

Kirfel A, Lippmann T (2002) Electron density distribution in fayalite, Fe_2SiO_4, a high-energy synchrotron radiation study, http://hasyweb.desy.de/science/annual_reports/2002_report/part1/contrib/42/6940.pdf.

Kittel C (1980) *Einführung in die Festkörperphysik*, Oldenbourg-Verlag, München-Wien, p. 534.

Klockmann F (1942) *Klockmann's Lehrbuch der Mineralogie* (Ramdohr P, ed), F. Enke Verlag, Stuttgart.

Kündig W, Cape JA, Lindquist RH, Constabaris G (1967) Some magnetic properties of Fe_2SiO_4 from 4 to 300 K, *J Appl Phys*, **38**(3), 947–948.

Lauer S, Marathe VR, Trautwein AX (1979) Sternheimer shielding using various approximations, *Phys Rev*, (A)**19**, 1852–1861.

Lottermoser W (2015) *Festkörperphysik—Die Wichtigsten Themen in 9 Kapiteln*, bookboon.com. Auflage, London, pp. 1–181.

Lottermoser W, Forcher K, Amthauer G, Treutmann W, Hosoya S (1996) Single crystal Mössbauer spectroscopy on the three principal sections of a synthetic fayalite sample in the antiferromagnetic state, *Phys Chem Minerals*, **23**, 432.

Lottermoser W, Fuess H (1992) A Polarized Neutron Study on Synthetic Mn_2SiO_4, *Phys Chem Minerals*, **19**, 46–51.

Lottermoser W, Schaper AK, Treutmann W, Redhammer G, Tippelt G, Lichtenberger A, Weber SU, Amthauer G (2006) ^{57}Fe Mössbauer spectroscopy on multiwalled carbon nanotubes with metal filling, *J Phys Chem B*, **110**(20), 9768–9771.

Lottermoser W, Steiner K, Grodzicki M, Jiang K, Scharfetter G, Bats JW, Redhammer G, Treutmann W, Hosoya S, Amthauer G (2002) The electric field gradient in synthetic fayalite α-Fe_2SiO_4 at moderate temperatures, *Phys Chem Minerals*, **29**, 112–121.

Lottermoser W, Weber SU, Grodzicki M, Kirfel A, Amthauer G (2012) A semi-quantitative approach to derive the electric field gradient, applied to synthetic fayalite, α-Fe_2SiO_4: A reappraisal, *Eur J Mineral*, **24**, 791–797.

Malmfors KC (1955) Resonant scattering of gamma-rays, in *Beta and Gamma Ray Spectroscopy* (Siegbahn K, ed), North Holland, Amsterdam.

Marshall W, Lovesey SW (1971) *Theory of Thermal Neutron Scattering*, Clarendon Press, Oxford.

Moon PB (1951) Resonant nuclear scattering of gamma-rays: Theory and preliminary experiments, *Proc Phys Soc London*, **A64**, 76.

Mössbauer Effect Data Center (MEDC), Dalian Institute of Chemical Physics (DICP), Chinese Academy of Sciences, 457 Zhongshan Road, Dalian, 116023, China, www.medc.dicp.ac.cn.

Müller R (1985) Magnetische Eigenschaften von Kobaltvanadat und Eisensilikat, Thesis University of Frankfurt, Frankfurt/M.

Müller R, Fuess H, Brown PJ (1982) Magnetic properties of synthetic fayalite, *J Phys C7*, **43**, 249–252.

Nomura S, Santoro R, Fang J, Newnham R (1964) Antiferromagntism in cobalt orthosilicate, *J Phys Chem Sol*, **25**, 901–905.

Phywe XR4.0 X-ray expert unit, Betriebsanleitung des Röntgengeräts 35W 09057-99, Phywe Systeme GmbH und Co. KG, Göttingen.

Pound RV, Rebka GA Jr. (1959) Gravitational red-shift in nuclear resonance, *Phys Rev Lett*, **3**(9), 439–441.

Research Systems Inc. (2002), IDL Interactive Data Language Ver. 5.6 and 6.3, Boulder, Colorado.

Robie RA, Finch CB, Hemiingway BS (1982) Heat capacity and entropy of fayalite (Fe_2SiO_4) between 5.1 and 383K: comparison of calorimetric and equilibrium values for the QFM buffer reaction, *Am Miner*, **67**, 463–469.

Santoro RP, Newnham RE, Nomura S (1966) Magnetic properties of Mn_2SiO_4 and Fe_2SiO_4, *J Phys Chem Sol*, **27**, 655–666.

Sheldrick G (1997) SHELX-97, program for the solution and refinement of crystal structures, University of Göttingen, Göttingen.

Slater JC (1974) *Quantum Theory of Molecules and Solids*, vol. 4, McGraw-Hill, New York.

Sternheimer RM (1963) Quadrupole antishielding factors of ions, *Phys Rev*, **130**, 1423–1425.

Stevens JG, Khasanov AM, Miller JW, Pollak H, Zhe Li (eds) (1998) *Mössbauer Mineral Handbook*, University of North Carolina at Asheville, North Carolina.

Stewart RF, Speckman MA, Flensburg C (1996) VALRAY 96, Valray programme package, Copenhagen.

Stout GH, Jensen LH (1989) *X-Ray Structure Determination: A Practical Guide*, 2nd ed, Wiley, West Sussex, UK.

Streltsov VA, Belokoneva EL, Tsirelson VG, Hansen NK (1993) Multipole analysis of the electron density in triphylite, $LiFePO_4$, using X-ray diffraction data, *Acta Cryst*, **B49**, 147–153.

Stummel F, Hainer K (1971) Praktische Mathematik, B. G. Teubner, Stuttgart.

Wagner U, Wagner FE, Häusler W, Shimada I (2000) The Use of Mössbauer Spectroscopy in *Studies of Archaeological Ceramics in Radiation in Art and Archaeometry* (Creagh DE, Bradley DA eds), Elsevier, Amsterdam.

Watson RE, Freeman AJ (1961) Hartree-Fock atomic scattering factors for the iron transition series, *Acta Cryst*, **14**, 27–37.

Watson R, Freeman AJ (1961) Origin of effective fields in magnetic materials, *Phys Rev*, **123**, 2027.

Weber S (1997) JAVA Lauegram: jcrystal.com/steffenweber/JAVA/jlaue/jlaue.html.

Zachariasen WH (1967) A general theory of X-ray diffraction in crystals, *Acta Cryst,* **23**, 558–564.

Zherebetskyy D (2010) Quantum mechanical first principles calculations of the electronic and magnetic structure of Fe-bearing rock-forming silicates, thesis at Faculty of Natural Sciences, University of Salzburg, Salzburg.

Index